M

Fro
mo
live

edite
Moh

facet publishing

Published by Facet Publishing,
7 Ridgmount Street, London WC1E 7AE
www.facetpublishing.co.uk

Facet Publishing is wholly owned by CILIP: the Chartered Institute of Library and Information Professionals.

British Library Cataloguing in Publication Data
A catalogue record for this book is available from the British Library.

ISBN 978-1-85604-944-3

First published 2013

FSC
Mixed Sources
Product group from well-managed forests and other controlled sources
www.fsc.org Cert no. SA-COC-1565
© 1996 Forest Stewardship Council

Text printed on FSC accredited material.

Typeset from editors' files by Flagholme Publishing Services in 12/14 American Garamond and Nimbus Sans.
Printed and made in Great Britain by CPI Group (UK) Ltd, Croydon, CR0 4YY.

Contents

Acknowledgements

Thanks to the authors who have contributed to the book and to those who have helped to compile the book. Special thanks to Christina Sutcliffe, who worked hard and diligently to help put the book together. Many thanks to the Open University of the United Kingdom, which hosted the M-Libraries 4 Conference and to Athabasca University, which helped to co-ordinate the conference. Thanks to the International Organizing Committee, Programme Committee and Local Organizing Committee, all the conference speakers, session chairs, delegates, helpers and sponsors for making the M-Libraries 4 Conference a success.

Every effort has been made to contact the holders of copyright material reproduced in this text, and thanks are due to them for permission to reproduce the material indicated. If there are any queries please contact the publisher.

Mohamed Ally
Gill Needham

Contributors

Keynote/invited speakers

Parveen Babbar is Assistant Librarian in Library and Documentation Division at Indira Gandhi National Open University in New Delhi, India. Prior to this he had been associated with the University of Delhi, India and has more than ten years of professional experience. He holds an MPhil (Library and Information Science), MLISc, Masters of Computer Application and Masters of Business Administration. He has been a Gold Medallist in MLISc from Delhi University, India and a Junior Research Fellow. He has received many awards and scholarships, including the SLA Asian Librarian Award at the SLA 2010 Annual Conference in New Orleans, USA.

Char Booth is an early riser, devoted oceanite and advocate of radical neutrality. She explores the integration of education, research, technology and design in libraries. Char is the Instruction Services Manager and E-Learning Librarian at the Claremont Colleges, and is on the faculty of the ACRL Information Literacy Immersion Institute. An ALA Emerging Leader and Library Journal Mover and Shaker, she blogs at *info-mational* and tweets *@charbooth*. Her publications include *Reflective Teaching, Effective Learning: Instructional Literacy for Library Educators* (ALA Editions, 2011) and *Informing Innovation: Tracking Student Interest in Emerging Library Technologies* (ACRL, 2009). Char received a BA in History from Reed College in 2001; an MSIS from the University of Texas Austin School of

Information in 2005, and a Master of Computer Education and Technology from Ohio University in 2008.

Bob Gann is Head of Strategy and Engagement for NHS Choices at the Department of Health. He joined the Department from NHS Direct, where he was Director of New Media, responsible for the NHS Direct website and digital TV services. In his career Bob has worked in healthcare libraries, in NHS public affairs and as chief executive of a not-for-profit agency providing health information services. He has also been a writer and editor for the British Medical Association, and had a regular column in *British Medical Journal*, 'What your patients are reading'. Bob has served on a number of working parties and task forces and was one of the 25 NHS leaders who signed the NHS Plan. He is Visiting Professor in Health Informatics at Plymouth University.

Contributors

Jo Alcock, as both a librarian and keen user of mobile technologies, has been following, experimenting and reporting on the area of mobile technologies in libraries since getting an iPod Touch in 2007. She blogs at Joeyanne Libraryanne (www.joeyanne.co.uk) and tweets as *@joeyanne*.

Professor Mohamed Ally is Professor and Researcher in the Technology Enhanced Knowledge Research Institute (TEKRI) at Athabasca University. He obtained his PhD from the University of Alberta, Canada. His current areas of research include mobile learning, e-learning, distance education and use of information and communication technology in training and education. Dr Ally is Past President of the International Federation of Training and Development Organizations (IFTDO) and is one of the Founding Directors of the International Association of Mobile Learning (IamLearn). He has recently edited four books on the use of mobile technology in education, training and libraries.

John Paul Anbu K is currently Head of Periodicals at the University of Swaziland Libraries. His contributions include an open source circulation control module for UNESCO's CDS-ISIS, an information system on 'Tea Disease Management' and a number of automation and digitizing projects. Being a regular invited speaker at a number of International Conferences, he has contributed a number of research articles to international journals and has co-edited two conference volumes. He is also Swaziland's country co-ordinator for eIFL-FOSS and INASP.

Bukky Olufemi Asubiojo is the University Librarian of Hezekiah Oluwasanmi Library, Obafemi Awolowo University, Ile-Ife, Osun State, Nigeria. She obtained both her Bachelor's (1983) and Master's degrees in Library and Information Science (1985) from the University of Ibadan and postgraduate diploma in Computer Science from Obafemi Awolowo University in Nigeria (2003).

Maneesh Kumar Bajpai, PhD has been an assistant librarian and librarian in charge of the Library of Dr RML National Law University, Lucknow, Uttar Pradesh, India since 2006. He holds an MLISc, PGDCA, BSc, PhD and UGC-NET certificate. His areas of interest are digital library and advanced library services and Web 2.0. He is also Executive Editor (Technical) of the journal *The Legal Analyst*. Dr Bajpai has work experience with organizations such as DRDO, etc. He has presented and contributed papers to national/international journals, conferences and seminars.

Matthew Borg is a librarian in the Virtual Learning Centre at Sheffield Hallam University. He looks after the design, look and feel of the library's online presence. He is interested in how we can create simple user interactions so that they can help resource discovery for library users. He is particularly interested in library website design, how we teach information literacy in a web-discovery world and knowledge transfer between information professionals.

Stephen A. Bowman is Deputy Librarian at the University of Chichester. A Fellow of CILIP, and a Fellow of the Higher Education Academy (HEA), he has an MA in Information Management, and an MSc in E-Learning. His interests include the history of the book, film and cinema, and Prog Rock.

Angels Carles-Pomar has worked in the Library Services of Universitat Autònoma de Barcelona (UAB) since 1985. She is currently Head of UAB Medical Library. Her interests are in library support services, user training, literacy and e-learning. Angels has a PhD and holds degrees in Librarianship, History and Documentation. She is also interested in political and social history.

Ana Castellano López has worked in Computer Services at the Universitat Autònoma de Barcelona (UAB) since 1989. She is currently Head of the Learning Resources Centre of the Faculty of Medicine at UAB. She is interested in online education and virtual learning environments. She has degrees in Biology, Computer Engineering and Psychology.

Dr Seema Chandhok is Deputy Librarian at Indira Gandhi National Open University (IGNOU), New Delhi, India. She has more than 23 years' professional experience, including 18 years as Assistant Librarian in IGNOU. She has Master's degrees in Public Administration, Distance Education and Library and Information Science and a PhD in Library and Information Science. Seema has had several articles published in professional journals and conference proceedings and has also published indexes to articles published in journals.

Caroline Corneau is a website manager for the Library of the European Parliament and has been in charge of mobile services since the launch of the project in 2011. First trained as a linguist, she holds a Master's degree in Information Science and Technology. Her interests include usability and design of library websites, mobile strategy and technologies for library services and responsive design.

Kate Cushon is a liaison librarian at the University of Regina (Canada), responsible for collection development and instruction support for several faculties. Her professional interests include promoting information literacy, improving library accessibility for marginalized groups and the adoption of new technologies in academic libraries.

Pete Dalton, MBA CMC FIC is Director of Evidence Base at Birmingham City University. He has many years' experience in managing and conducting a wide range of research, evaluation and consultancy projects in the information and education sectors and beyond. He is a strong advocate of evidence-based decision making in policy and practice.

Neil Ford is Academic Liaison Development Manager for Library and Learning Support at Bournemouth University. He has previous experience in health, public, commercial and academic libraries. His current interests include digital literacy and academic skills support, QR codes, collection development and co-creation of reading lists in academic liaison.

Fernando Guerrero Rebollo has worked at Computer Services of Universitat Autònoma de Barcelona (UAB) since 1987. He is currently Head of UAB computer services and multimedia resources in the Faculty of Translation and Interpreting. His interests are in the Open Source Initiative and the use of video as a teaching resource for e-learning on mobile devices. Fernando has degrees in Documentation and Primary Education.

Dr Anne Hewling is an education consultant specializing in instructional design and research for development. She completed her PhD in educational technology at the Open University and worked there for several years on new technology projects, including PROWE, which concentrated on wikis and blogs to support staff development, and on iKnow, which uses web and mobile tools to improve workplace information literacy and management skills. She currently divides her time between designing and managing education projects in the developing world and teaching e-learning innovation and international development – mainly online.

David Honeybone is a liaison librarian at the Brownless Biomedical Library, University of Melbourne, Australia. His interest in social media developed while he was studying for his Master's degree. He has looked after the library Twitter account, @unilibrary, since June 2011, creating a popular interactive tool through a blend of helpful information and bad jokes.

Md. Anwarul Islam acquired BA Honours, MA and MPhil degrees in Information Science and Library Management from Dhaka University, Bangladesh. He is currently an Assistant Professor and student adviser in the same department. Before that, he worked in GrameenPhone. His research areas are ICT, information literacy and digital libraries.

Dr Sridevi Jetty, deputy Librarian at Bundelkhand University, Jhansi, India is an information specialist with two decades of experience in library and information science. She received the Commonwealth Professional Fellowship from Middlesex University, London, UK in 2008. She has to her credit 19 research papers and articles published and presented in various national and international forums.

Keiso Katsura is Professor of Library and Information Science at Miyagi Gakuin Women's University, Sendai City, Miyagi, Japan. He was educated at Keio University, Tokyo (BA Library and Information Science, 1973; MA Sociology, 1977) and Columbia University (MS Library Service, 1988). He worked as a librarian and information professional at the Japan International Cooperation Agency, Tokyo from 1979 to 1992. He is currently involved in university courses, delivering web and mobile-based lectures with a range of Web 2.0 content. His main area of interest is in the information search and retrieval literacy of academics.

Peter Kilroy is a freelance lecturer in Cultural Studies with an interest in web design, e-learning, mobile learning and early recording technologies. He is based at the Faculty of Performance, Visual Arts and Communications at Leeds University and at Leeds University Library.

Bianca Kramer holds a PhD in Neurobiology and works at Utrecht University Library as an information specialist in health and medical sciences. She is involved in developing e-learning, teaching information literacy and consulting on search strategies for systematic reviews and meta-analyses. She has set up in-house training programs for Web 2.0 and social media.

Sa'ad Laws is an information and reference service librarian at Weill Cornell Medical College in Qatar. His main duties involve working with medical faculty and students. Most recently he has helped to implement a mobile device program. Sa'ad graduated from San Jose State University with an MLIS and from George Mason University with an MA in Anthropology.

Lili Luo is an assistant professor at the School of Library and Information Science at San Jose State University. Her research focuses on information access and services in the digital age, library and information science education, research methods, evaluation of information services and virtual librarianship.

Binky Lush is a web developer/designer in the Department for Information Technologies at the Penn State University Libraries, where she co-ordinates the design and development of the libraries' website. Binky also co-leads the libraries' Discovery and Access Team, which is focused on enhancing the libraries' discovery tools and ensuring an engaging, unified and user-centred search and discovery experience. Binky's professional interests include the exploration and application of emerging technologies in the library.

Keren Mills is a librarian in a research and innovation role and has been a part of the team developing mobile library services for Open University students since 2006. Keren had the opportunity to explore mobile user requirements through the Arcadia fellowship at Cambridge University in 2009. She is now responsible for the content and usability of the Library Services website and has a role in developing and commissioning rich media content. Most recently she has been managing the 'Mobilising Academic Content Online' (MACON) project, which is developing a mobile search tool using the EBSCO EDS API.

Kay Munro is convenor of the University of Glasgow Library's Mobile Technologies Group. Kay is a College Librarian providing support to the College of Social Science.

Gill Needham is currently Associate Director (Information Management and Innovation) at the Open University Library. Her varied career has encompassed roles as researcher, librarian, project manager and educator. She has more than 25 years' experience in senior roles in the library and information sector, with an emphasis on innovation and skills development. Since joining the Open University she has led strategic work in areas such as information literacy, flexible learner support and library systems development and has authored a number of Open University courses. She was awarded a National Teaching Fellowship by the Higher Education Academy in 2006. Her current work interests are digital literacy, mobile developments and digital libraries.

Georgina Parsons is currently Systems Librarian at Brunel University, having previously worked with a library management system supplier. As her responsibilities cover a variety of library systems, she is interested in integration of these and innovations in systems and service delivery, especially for mobile.

Emily Rimland is an Information Literacy Librarian and the Learning Technologies Coordinator at the Pennsylvania State University, where she enjoys providing instruction, reference and outreach services to undergraduate students. She holds an MLIS from the University of Pittsburgh and her research interests include the application of emerging technologies to library services, information literacy and instructional technologies.

Louise Saults, currently working at National University of Ireland Maynooth, has previously worked in various departments in the Library, in more recent years in Collection Management Services. She has worked on a number of projects, one of which was to make core text e-books available to students. In conjunction with the Departments of Modern History and Business and Law she has completed a pilot mobile technology project which involved making Kindles that contained their course reading available to students.

Hassan Sheikh works for Sidra Medical and Research Center in Qatar as the Technical Lead for Library systems. Previously Hassan was the

Head of Systems Development at the Open University (OU) (UK) Library. His previous role was the technical lead on several internal and external projects, including the Library website, development of mobile library services, access and authentication, and the Library Systems Strategy. Hassan has several years' experience in programming and usability evaluations and worked in the Institute of Educational Technology at the OU for four and a half years before joining Library Services in October 2006.

Rosemary Stenson is a member of the University of Glasgow Library's Mobile Technologies Group. Rosemary is Head of Information Resources Services, with particular responsibility for cataloguing policy.

Karen Stevenson is a member of the University of Glasgow Library's Mobile Technologies Group. Karen is the Library Management System Administrator.

Wendy Walker is a member of the University of Glasgow Library's Mobile Technologies Group. She is Head of Metadata Services and her responsibilities include the management and promotion of the library's non-serial e-resources collections.

Rowan Williamson has worked in academic libraries for the past ten years, in both further and higher education. She has been the Learning Resources Manager for the Art, Design and Architecture campus at Kingston University for the past five years. Rowan has worked on a number of projects to develop services and new ways of working which incorporate new technologies in response to how students prefer to work, including using iPads for roving, implementing room and equipment booking systems and designing online tutorials for information skills. She is an active member of the profession and a committee member for the Academic and Research Libraries Group, London and the South East group.

Andrew Wilson is the Senior IT Advisor for Information Services at the University of Stirling. He is the creator and lead developer of its mobile platform, mStir, a device-agnostic mobile web app. Andrew has a background in web development and has been involved in large projects such as the James Hogg Research Project and the Scottish Centre for Crime and Justice Research. He is also interested in promoting information services departments through creative and innovative approaches.

Foreword

It should not come as a surprise that library mobility is no longer a question: rather, it is firmly a reality. Fellow M-Libraries 2012 keynote speaker and UNESCO Programme Specialist in Mobile Learning Steve Vosloo noted that, as of 2012, there are close to 6 billion mobile subscriptions worldwide, 77% from developing nations alone.[1] In January 2012, the Pew Internet and American Life project reported that approximately 20% of Americans owned a dedicated e-reader.[2] Web-enabled devices now facilitate ground-breaking literacy and learning projects and often provide the sole means of connectivity in low-resource areas. For all of these reasons and more, libraries and librarians worldwide must examine their orientation to the new mobile reality.

The lightning rise of mobile technology means that users' expectations of mobile-optimized content and applications have become a foregone conclusion. Which is not to say that mobile tools, content, sites and apps should be taken up without prior critical consideration of user needs and available resources. As much as mobility can (and should) become a guiding force in library technology development, it is imperative that we also embrace *complementarity*, or the recognition that a range of formats, media and devices have equally important roles in the lives and preferences of users. Libraries have always celebrated a plurality of information sources and access methods, and the rise of mobile tools and e-reading represent yet another phase in our tirelessly shifting information context.

To truly innovate in mobile access and learning, libraries must not only embrace their role in the complementarity revolution, they must share what they learn from their model projects and their user experiences in order to push the field forward. If mobile access, research and reading are to become seamless in libraries, informed design and transparent communication about libraries' collective efforts will be key. Sharing the impact and implications of mobile technology in libraries is the focus of the annual M-Libraries Conference and of the excellent collection of papers you will find herein.

Contributors to this volume showcase a diverse landscape of mobile development and deployment, from device-specific projects to content optimization and mobile learning programmes. These early adopters communicate the type of strategic insight that will be crucial to a more collectively informed library mobility. When we learn from the advances and pitfalls of our peers through forward-thinking forums such as M-Libraries, our progress toward a better user experience is assured.

Char Booth

References

1 Vosloo, S., *The Power of Mobile to Educate, Support and Engage: examples from around the world*, M-Libraries 2012 keynote, www.m-libraries.org/news/online-sessions.

2 Rainie, L., *Tablet and E-book Reader Ownership Nearly Double Over the Holiday Gift-Giving Period*, Pew Internet and American Life Project, 23 January 2012, http://libraries.pewinternet.org/2012/01/23/tablet-and-e-book-reader-ownership-nearly-double-over-the-holiday-gift-giving-period/.

Introduction

Information consumption is and will increasingly be on the move because of the proliferation of emerging technologies to access information. People have the technology to access information from anywhere and anytime, which requires libraries to provide the services to meet their expectations. This book contributes to the expansion of information on the move using mobile technology by showcasing successful implementation and best practice to inform and inspire others. In some instances, libraries have to be transformed in order to take up the challenge of providing services to mobile users. The chapters in Part 1, Transformation, illustrate how libraries and librarians have been transformed and how this can be achieved, with services ranging from SMS messaging for content alerts to enhancing access to collections using QR codes. Part 2, Inspiration, is a collection of chapters selected to motivate others and inspire reflection and discussion. Examples are given of the power of mobile technologies to support mass education in India and to challenge the traditional delivery of healthcare in the UK and worldwide, alongside an eclectic collection of thought-provoking contributions from educators and librarians. In Part 3, Implementation, library practitioners from around the world describe how they have planned and implemented highly successful and innovative mobile services.

The chapters in this book are authored by experts, practitioners and researchers from around the world, providing a global perspective and picture of the use of mobile technology in libraries. As more library resources and learning materials become available in electronic format, it is crucial for users to be able to access them using the mobile device of their choice. Librarians have an important role to play in facilitating users' access to quality electronic information and learning materials and supporting users who need help with using the technology. This is increasingly important in lifelong learning, where people of different age groups are accessing online information and may not have the option of using print.

To conclude, the publication of this book is timely in order to accelerate the use of mobile technology for the consumption of information on the move. In the emerging technology age services are provided to people anywhere and anytime using a range of technologies. Information must be available for access by people from anywhere and at anytime using a variety of technologies. Librarians have a major role in facilitating access and providing the services required in order to access quality information.

Professor Mohamed Ally
Athabasca University
Canada

Part 1

Transformation

1

Cost-effective content alert system using SMS: a case study at Bundelkhand University Library, Jhansi

Sridevi Jetty, Maneesh Kumar Bajpai and John Paul Anbu K

Introduction

The past decade has seen a tremendous growth in mobile telephony. Research carried out by Infonetics predicts that by 2013 the number of worldwide mobile subscribers will reach 5.9 billion. This growth is ascribed mainly to the anticipated increase in mobile communication in Africa, India and China. Innovations in handsets and modes of transmission have seen an expansion of mobile telephone services to all walks of life. McEwen observes that Net Gen and Millennial students are used to being always with and on their mobile phones (McEwen, 2012).

Among the various mobile applications, Short Message Service (SMS) is one of the most popular services, widely used all over the world. Research shows that SMS has been adopted by many users and has in fact become extremely popular among both younger and older generations of users. It is estimated that close to 3.3 trillion SMS have been sent over the mobile network since its inception. Though some notable studies on the usage of SMS have been done in the past, there is very little information available on the connection between SMS and library services, although it can be generally noted many library services can be successfully run through SMS services. Bundelkhand University, in Jhansi in the heartland of India, attempted to use SMS technology to implement a cost-effective content alert system. This paper is a product of that project, which looked into whether such a cost-effective system could be implemented using SMS technology, its implications for the use of library resources and the acceptance of SMS services amongst library users.

Content alerts

An alert is defined as a form of communication that is important and time sensitive and that enables the recipient to be alert to issues which are related to them. In the library world an alert contains user-requested information or content. Alert messaging or alert notification is thus the delivery of alerts to recipients. Most sophisticated service providers embrace all capabilities, aggregating a multitude of reminders, notifications and alerts and adapting the delivery system to the specific context of the content being delivered, thus enabling users to create sophisticated scenarios of their own choosing.[1]

In the academic world alerts contain the content or a reference to the content which the user has requested. Most academic publishers and database aggregators provide content alerts to their electronic databases as a value-added service which is relevant to the content and context. Commonly, these alerts are delivered to the user's mailbox through e-mail systems. At Bundelkhand University an attempt was made to convert these e-mail alerts into a more time-sensitive SMS-based value-added service to library users, so as to maximize the usage of library resources.

It is worth observing that mobile applications for libraries are in their infancy. A number of Integrated Library Systems (ILS) have started to adopt mobile components for their OPACs; some ILS provide various other mobile-related services, especially alert messages using SMS. Libraries use SMS to interact with their members and to send alerts and notifications as a general value-added service. In developed countries libraries are using SMS to interact with their members and send alerts and notifications related to the library. Timothy Vollmer observes that mobile technology is changing the relationship between libraries and their users by expanding services and posing new challenges to readers' privacy (Vollmer, 2010).

Literature review

There is growing interest within academic institutions in using mobile devices to support teaching and learning. A variety of mobile devices can be used in mobile learning. The most ubiquitous and stable mobile technology, namely Short Message Service (SMS) or texting (Traxler, 2005) on mobile phones, has great potential in education. Over the last 10 years, many SMS projects for teaching and learning have been reported in the literature. Library services can be improved through SMS-based administrative support. Libraries can reach out and serve students ubiquitously by sending and receiving SMS-based library information. There are a number of areas in library services where SMS-based messages can be helpful. Basic information alerts such as notices of book reservations, renewals and overdue reminders work well via this medium of communication. Walsh (2009) points to a number of pilot projects in the employment of SMS in the mainstream educational system.

One example is the SMS alert services offered by the Hong Kong Institute of Education (HKIED, 2012). Further library services can also be provided via SMS-based systems. For example, extended text messaging reference can text SMS messages to and receive answers from librarians, as reported in Hill, Madarash Hill and Sherman's (2007) research at Southeastern Louisiana University, as a way to further enhance the quality of services provided by libraries in higher education.

SMS has become a popular way of communicating, particularly amongst the X and Y generations, who make up a large proportion of our university client base. As a form of communication it is mobile, asynchronous and in everybody's hands. Although it is a widely used communication mechanism for mobile phone users, SMS is far more than just a technology for teenage chat. This is mainly because the technology is ubiquitous and customers can make calls anywhere to transact various kinds of business.

Custom-made technology such as Liblet (a Portalify product), AltaRama's Reference by SMS or Mosio's Text A Librarian allow clients to use SMS to send reference queries to the library; renew library books; pay library fines; and check the availability of library items. Liblet also enables the library to send due date reminders, recall and hold notices via SMS. Popular in Scandinavia, Liblet has been successfully implemented by Chalmers University of Technology Library in Sweden and by Helsinki University of Technology (Pasanen and Muhonen, 2002).

SMS technology

SMS technology evolved out of the Global System for Mobile Communications standard, an internationally accepted mobile phone network specification created by the European Telecommunications Standards Institute. The 3rd Generation Partnership Project maintains the SMS standard (Brown, 2007). SMS messages are handled via a short message service centre (SMSC) that the mobile provider maintains for the end devices. The SMSC can send SMS messages to the end device using a maximum payload of 140 octets. This defines the upper limit of an SMS message as 160 characters using 7-bit encoding. It is possible to specify other schemes such as 8-bit or 16-bit encoding, which decreases the maximum message length to 140 and 70 characters, respectively.

Text messaging is still the only universal mobile platform for the masses. It does not require special downloads, as it is already available on 98% of all mobile phones. As already noted, mobile networks allow text messages of up to 160 characters to be sent or received. The concatenation mechanism of SMS also permits messages of longer than 140 bytes. For a large operation to send and receive SMS messages, the messages can be sent to the mobile network operator using SMPP (Short Message Peer-to-peer Protocol). SMPP is an open industry standard messaging protocol that is designed to simplify the integration of data applications with wireless mobile networks. Most service providers offer e-mail to SMS gateways that receive e-mail messages and convert them into SMS to be sent to the mobile network.

Project overview

The project to set up an SMS service was comprised of the following steps:

- Choose the databases for which content alerts are to be received.
- Prepare a list of users with relevant subject fields for content harvesting.
- Create a user profile from this list.
- Create a separate mail ID for the project.
- Start e-mail content alerts from the chosen databases.
- On receipt of e-mail alerts send them to the users through SMS.
- Conduct a post-project review amongst the participants.

Bundelkhand University subscribes to the UGC Infonet Digital Library Consortium, INFLIBNET Centre Ahmedabad.[2] Almost all the databases were needed to cover the subjects that were present in the list of users, and so all the 19 databases were chosen. These were: American Chemical Society, American Institute of Physics, American Physical Society, Annual Reviews, Cambridge University Press, Economic and Political Weekly, Emerald, Indian Journals.com, Institute of Physics, ISID, JCCC, JSTOR, Oxford University Press, Project Muse, Royal Society of Chemistry, Springer Link, Taylor & Francis, Web of Science and Wiley-Blackwell. Then:

- A profile database of 50 users was created as an MS Excel spreadsheet, with the users' name, subject, subject keywords, e-mail ID and Mobile number (Figure 1.1).
- All of the users' keywords were registered with the database(s) of the UGC Infonet Digital Library Consortium.

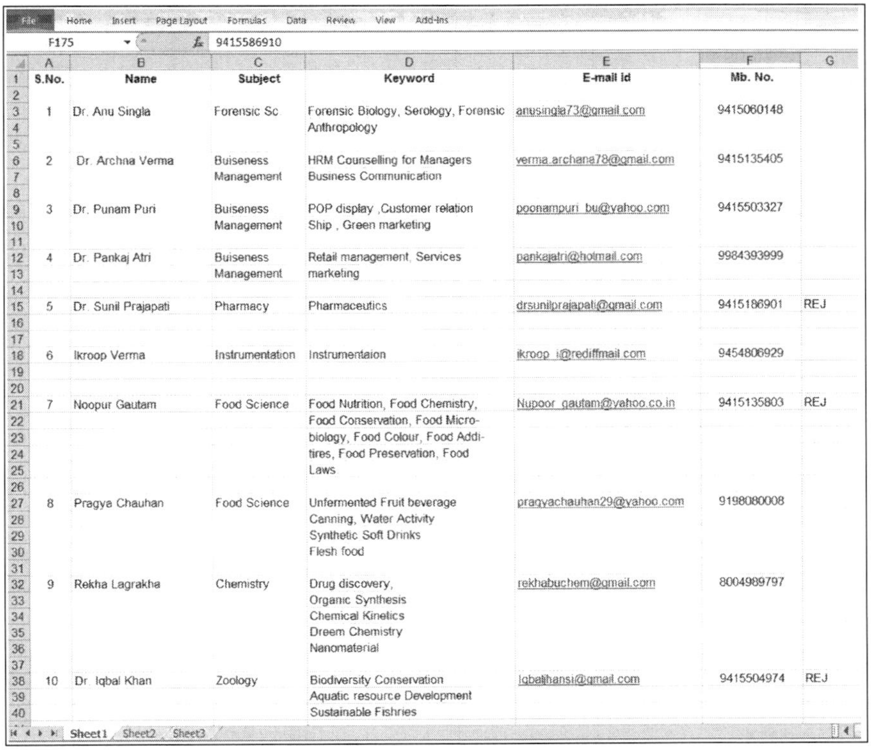

S.No.	Name	Subject	Keyword	E-mail id	Mb. No.	
1	Dr. Anu Singla	Forensic Sc	Forensic Biology, Serology, Forensic Anthropology	anusingla73@gmail.com	9415060148	
2	Dr. Archna Verma	Buiseness Management	HRM Counselling for Managers Business Communication	verma.archana78@gmail.com	9415135405	
3	Dr. Punam Puri	Buiseness Management	POP display ,Customer relation Ship , Green marketing	poonampuri_bu@yahoo.com	9415503327	
4	Dr. Pankaj Atri	Buiseness Management	Retail management, Services marketing	pankajatri@hotmail.com	9984393999	
5	Dr. Sunil Prajapati	Pharmacy	Pharmaceutics	drsunilprajapati@gmail.com	9415186901	REJ
6	Ikroop Verma	Instrumentation	Instrumentaion	ikroop_i@rediffmail.com	9454806929	
7	Noopur Gautam	Food Science	Food Nutrition, Food Chemistry, Food Conservation, Food Micro-biology, Food Colour, Food Addi-tires, Food Preservation, Food Laws	Nupoor_gautam@yahoo.co.in	9415135803	REJ
8	Pragya Chauhan	Food Science	Unfermented Fruit beverage Canning, Water Activity Synthetic Soft Drinks Flesh food	pragyachauhan29@yahoo.com	9198080008	
9	Rekha Lagrakha	Chemistry	Drug discovery, Organic Synthesis Chemical Kinetics Dreem Chemistry Nanomaterial	rekhabuchem@gmail.com	8004989797	
10	Dr. Iqbal Khan	Zoology	Biodiversity Conservation Aquatic resource Development Sustainable Fishries	Iqbaljhansi@gmail.com	9415504974	REJ

Figure 1.1 User profiles database

- The Library created a separate e-mail ID (centrallibrarybu@yahoo.com) for receiving the content alerts from all of the database(s).
- The Library hired a bulk SMS server (http://bulksms.tornadowebsolution.com) for sending SMS alerts at a very economical cost of Rs. 0.23/SMS (Figure 1.2).

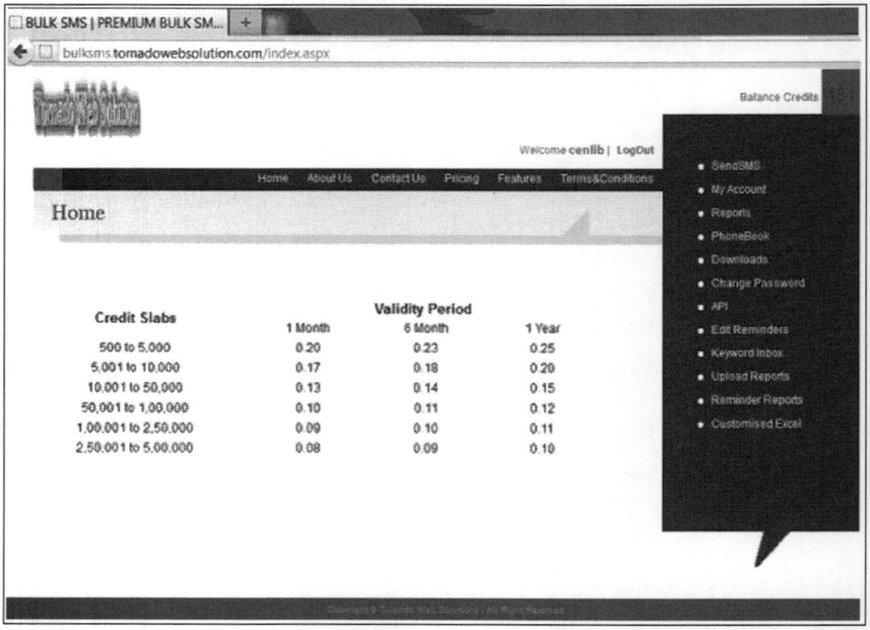

Figure 1.2 Bulk SMS server home page (reproduced with permission)

- On receipt of content alert messages from the database(s), these were transmitted to the specific users after cross-checking the profile database. This was done manually through the SMS gateway service, as shown in Figure 1.3.

Figure 1.4 illustrates an SMS alert received on a user's handset. A total of 600 SMS texts have been sent to users and the project is ongoing.

Findings

To gauge the usefulness of the content alert service using SMS, a survey

Figure 1.3 Onward transmission of SMS message, using the SMS gateway service (reproduced with permission)

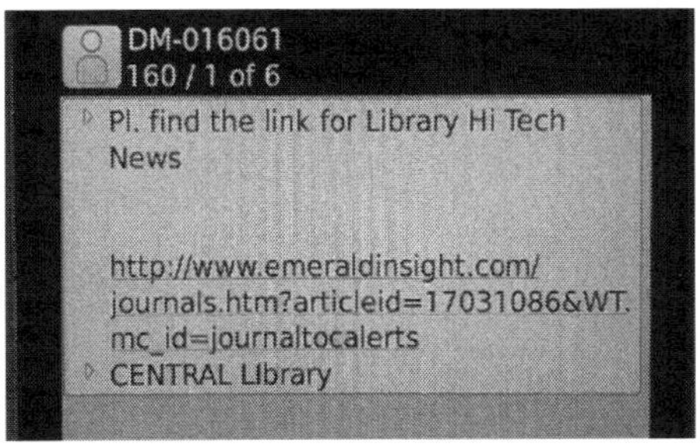

Figure 1.4 SMS alert received on a user's handset

was conducted six months into the project. A well-tested questionnaire was distributed among 36 (faculty members) out of the total 50 users. Of these, 34 responded and 2 did not. This equates to feedback from 68% of users, which, keeping in view the 10% standard sampling, is a greater response.

Close to 94% of respondents confirmed that they were well aware of the library's SMS alert service, which shows that they use their electronic devices frequently. Close to 41% confirmed that they received the SMS alert service daily; 35% received the service once a week; and 24% received it monthly. Given the frequency with which the library receives SMS alerts from the publishers, the frequency with which users received SMS alerts was very impressive.

Purpose of using SMS alerts

With a high percentage of users confirming their awareness of SMS alerts and an above average number receiving SMS alerts on a daily basis, users were asked the purposes for which they were using SMS alerts. Close to 71% responded that they used the SMS service for teaching; 76% for research; followed by 9% for scholarly writing. This provides a picture of the various activities for which SMS alerts were used.

Frequency of getting desired documents through SMS alert service

Users were asked how frequently they received the desired documents. Close to 50% of users reported that they were receiving 1–5 times the desired information through the SMS, whereas only 5% of users were getting such service frequently, i.e. more than 15 times. The variations in frequency may be due to the frequency of different publications.

Satisfaction with SMS alert service

Close to 79% of users responded that they were satisfied with the Library's SMS alert service, which reflects the success of the Library's efforts. This question was important to the Library administration for the continuation and further enhancing of the project.

Requirement of SMS alerts for other services

In order to broaden the SMS service, users were asked what other alerts they would like. A new book titles alert service was requested by 82%

of users, and close to 20% requested circulation alerts. This has given a direction for the Library's future enhancement of the SMS alert service.

Decision on the continuation of the service

When asked whether they wanted the project to continue, 97% of the surveyed users responded positively, which reflects the great acceptance of the SMS alert service. This has allowed the Library management to seek approval from the university administrators for the extension and continuation of the project at the university level.

Comfortable service opinion

Since SMS is a new medium of communication, the Library also wanted to know what kind of alerts the users were comfortable with. The majority of users (74%) were of the view that SMS alert service is more convenient than e-mail alerts, because in this mobile era everyone has a mobile phone to hand, whereas for e-mail access one has to have an internet connection.

Cost-effectiveness of SMS content alert service vs print/traditional methods

Previously, the content alert service was provided by the traditional method of circulating printed copies of alerts. The printing of a single page at Central Library, Bundelkhand University costs Rs. 0.75. Sending e-mails was considered to be cost-effective, but it proved not to be time-effective, as computer access for checking e-mail was limited to either the office or the home. The printing and messenger costs together come to around Rs. 2.00 or more for printed alerts, whereas the SMS alert service costs just Rs. 0.23 per alert per user. The service is very time saving, accurate and cost-effective.

Conclusion

The library initiated the SMS alert service to provide article alerts in their subject areas to the user community at Bundelkhand University. The system was found to be very useful for sending such alerts. A survey among those users who are being provided with the service shows that the users are very much aware of the service and are also greatly convinced that it is very useful for their research and academic work. Most users are in favour of continuing the service. The majority of users support using the SMS alert service for new book arrivals, overdue book alerts, etc., which shows the usefulness of the service and its impact on users.

The results reflect a very positive response to the provision of library and information services. Mobile technology is a very economic communication medium in India and accessible to everyone. The SMS alert service is very cost-effective and is successful in reaching all users frequently or whenever required. The indications are that the service should be continued long term and that it will certainly increase reading interests among the user community.

Thus the mobile phone – now with significant computing power – is the primary internet connection – and the only one for a majority of the people across the world, providing information in a portable, well connected form at a relatively low price.

Notes

1 http://en.wikipedia.org/wiki/Alert_messaging.

2 www.inflibnet.ac.in/econ/eres.php?memID=63.

References

Brown, M. (2007) New Gadgets Worth Exploring, *Library Hi Tech News*, **8**, 10–13.

HKIED (2012) *Library SMS Alert Service, Hong Kong Institute of Education*, https://libsms.ied.edu.hk.

Hill, J. B., Madarash Hill, C. and Sherman, D. (2007) Text Messaging in an Academic Library: integrating SMS into digital reference, *The Reference Librarian*, 47 (1), 17–29.

McEwen, R. and Scheaffer, K. (2012) Orality in the Library: how mobile phones challenge our understanding of collaboration in hybridized information centers', *Library & Information Science Research*, 34, 92–8.

Pasanen, I. and Muhonen, A. (2002) *Library in Your Pocket*, www.iatul.org/doclibrary/public/conf_proceedings/2002/muhonen.pdf.

Traxler, J. (2005) Case Studies: introduction and overview. In Kukulska-Hulme, A. and Traxler, J. (eds), *Mobile Learning: a handbook for educators and trainers*, Routledge.

Vollmer, T. (2010) There's an App for That! Libraries and mobile technology: an introduction to public policy considerations, *Policy Brief* No. 3 (ALA Office for Technology Policy), 1–14.

Walsh, A. (2009) Text Messaging (SMS) and Libraries. *Library Hi Tech News*, **8**, 9–11.

Further reading

Cummings, J., Merrill, A. and Borrelli, S. (2010) The Use of Handheld Mobile Devices: their impact and implications for library services, *Library Hi Tech*, **88** (1), 22–40.

Davidson, G. and Dorner, D. (2009) Selection Criteria for Mobile Library Collections, *Collection Building*, **28** (2), 51–8.

Goh, T. T. and Liew, C. L. (2009) SMS-based Library Catalogue System: a preliminary investigation of user acceptance, *The Electronic Library*, **27** (3), 394–408.

Hahn, J. (2008) Mobile Learning for the Twenty-first Century Librarian, *Reference Services Review*, **36** (3), 272–88.

Herman, S. (2007) SMS Reference: keeping up with your clients, *The Electronic Library*, **25** (4), 401–8.

Jacobs, M. L. (2009) Libraries and the Mobile Revolution: remediation=relevance, *Reference Services Review*, **37** (3), 286–90.

Jensen, R. B. (2010) Optimizing Library Content for Mobile Phones, *Library Hi Tech News*, **2**, 6–9.

Karim, N. S. A., Darus, S. H. and Hussin, R. (2006) Mobile Phone Applications in Academic Library Services: a students' feedback survey, *Campus-Wide Information Systems*, **23** (1), 35–51.

Murray, L. Libraries 'Like To Move It, Move It', *Reference Services Review*, **38** (2), 233–49.

Olatokun, M. W. and Bodunwa, I. O. (2006) GSM Usage at the University of Ibadan, *The Electronic Library*, **24** (4), 530–47.

Smura, T., Kivi, A. and Töyli, J. (2009) A Framework for Analysing the Usage of Mobile Services. *Info*, **11** (4), 53–67.

Stahr, B. (2009) SMS Library Reference Service Options, *Library Hi Tech News*, **3** (4), 13–15.

2

From aspiration to innovation: the Live Lab concept at the University of Glasgow Library

Rosemary Stenson, Wendy Walker, Kay Munro and Karen Stevenson

Introduction

The Live Lab concept is one of the strands of the mobile strategy at the University of Glasgow Library. During the course of 2010 the Library's Mobile Technologies Group (MTG) developed the strategy in response to evidence from Google Analytics of steadily increasing traffic from mobile devices to the Library website. The academic literature also indicated that as ownership of smartphones and tablets increased there was a growing expectation from library users that services and resources should be capable of being delivered to mobile devices.

The mobile strategy

In planning the strategy[1] the intent was to design a framework for the investigation, development and support of mobile service delivery. The strategy established a comprehensive set of goals across a wide range of services and includes an annual work plan for the implementation phases. Each planning cycle covers an academic session. There are ten areas of strategic priority. These include an annual mobile survey of Library users; the development of a mobile website incorporating search, user account and access to e-resources functionality; digital media skills for Library staff and users; and the Live Lab.

Project management of each of the implementation phases is a central component of the strategic process and is critical to ensuring that targets and deadlines are met. The MTG is responsible for the co-ordination of the entire project and for determining specific priorities for each phase. A full review is carried out at the end of each phase and this, together with a review of current academic literature and the results from the annual mobile survey, is used to set targets for the next stage of the planning cycle.

Working groups, convened by a member of the MTG, are established to take forward each area of strategic activity. The groups are given specific remits and deadlines for delivery and are tasked with reporting progress to the MTG at its regular monthly meetings. Responsibility for developing the mobile website and the Live Lab elements of the strategy remains with the main group.

Several strategic targets relate directly to service delivery. This can include planning for completely new services, or the enhancement of a service introduced during an earlier phase. Other strands of the strategy seek to put in place the framework required to support all aspects of mobile delivery. Underpinning the whole strategy, and part of this support framework, is the concept of what has become known as the Live Lab.

The Live Lab concept

From the outset the MTG took the view that access to a variety of mobile devices, i.e. e-readers, phones and tablets, was essential. It envisaged three main purposes of the lab: familiarity and engagement with the technology; testing of services and content; and engaging Library staff in the development of mobile services.

It was vital that the members of the MTG gained expertise in the operation of the different devices, and evaluated their strengths and weaknesses and their potential for delivering services and accessing resources. In seeking to put the user at the centre of the strategy, the group also hoped to develop a better understanding of the user experience.

Equally important is the need to test. New services to be introduced require testing across the range of devices and operating systems. As one of the strategic priorities is mobile access to e-resources, testing of e-content and how it can be delivered and accessed via the various devices is also an essential requirement.

Engaging Library staff in the planning and delivery of mobile services is fundamental to the ongoing success of the strategic process. Allowing access to the purchased devices to the wider Library staff was seen as an opportunity to begin this process. Additionally, this would provide an opportunity to expose Library staff to the work being undertaken by the various mobile groups.

In accepting the strategy proposal, senior Library management allocated funding for the purchase of a range of mobile devices. The Live Lab had originally been envisaged as a 'virtual' lab; staff would use the devices for testing of services, or borrow them for use at work or for personal use at home. However Library management additionally allocated a room within the Library to house the lab. The space was turned into a relaxed, informal area with improved Wi-Fi and 3G connectivity. The Live Lab therefore became a physical entity as well as a virtual one, and having a permanent base presented opportunities for extending the function of the lab.

The Live Lab in action

With the purchase of the devices and the availability of the physical space, the Live Lab was brought into use initially by the various MTG working groups. The groups began to familiarize themselves with the devices and, depending on the function of each group, to test either new services or content. For example, the mobile website was developed using iWebKit5,[2] and testing of this software was done across the range of devices and the different operating systems. Testing of content was done by the E-resources Group. Initially this group looked at testing the Library's subscribed academic e-book content. This was less than successful. The problems of downloading this content onto mobile phones, tablets or e-readers are now well documented. For example, see Gibson and Gibbs (2011) on accessing e-content on e-readers. In later phases this group has moved on to investigating the area of e-journal and database content, where suppliers and aggregators are gradually making their content accessible via mobile devices. The availability of devices has been crucial to the success of this group, and links to some established database content are now available via the mobile website.

One of the most successful uses of the devices from the Live Lab has been in support of the annual mobile survey. The strategy is based on an assumption that the Library will provide services on user-owned devices. Evidence from Google Analytics and the Library's regular survey of its users regarding ownership of devices informs the decision on which devices to purchase. The survey is seen as vital to the planning of future services and in tracking changing user behaviour and demand, but, as with most surveys, there are often problems with participation rates. In the second annual survey completion rates were initially low. One proactive attempt to encourage more returns involved staff targeting users in the social areas of the Library. Armed with various tablets from the Live Lab, staff asked users to complete the survey on the devices. Users appeared to like the novelty of handling and using the devices, and this method helped to increase the response rate over the previously conducted survey by over 125%. It also marked the first attempt by the MTG to involve Library staff directly in planning for mobile services. Some basic training in how to use the various tablets was offered to the staff involved, and they were also given background on the mobile strategy and the new services which were due to be introduced.

Digital media skills gap

During the first phase of the strategic activity, the MTG hosted drop-in sessions for Library staff, both to preview the new services, e.g. the mobile library catalogue and QR codes, and to introduce the Live Lab itself. The sessions allowed staff a hands-on opportunity to try out the devices and provided an informal setting for asking questions. Although voluntary, the sessions were extremely well attended. Staff were enthusiastic and engaged and could see the potential of mobile to improve and enhance the library experience for users. Somewhat unexpectedly, it also prompted some to consider how this technology might be applied to current operational processes within their own departments. Nevertheless, it quickly became apparent that the staff who attended, despite their interest, were very unsure of the technology. They acknowledged that they lacked the expertise and familiarity with both the devices and the underlying infrastructure. This issue was apparent at all levels of staff and across the age ranges. It had been anticipated that younger staff members would be very familiar with the mobile environment. This was not the case, as most found the cost of the devices and associated tariffs unaffordable. For others, many had simply not engaged with the technology at that point.

A vital aim of the strategy is to develop the digital media skills of our users: 'In the future, we must be able to teach users about how to access information from their mobile devices and show them how to organise and use effectively that information for coursework and research' (University of Glasgow Library, 2010). The evidence from the Live Lab drop-in sessions resulted in a major shift in focus for the Digital Media Skills Group, which was set up to take this aspect of the strategy forward. The lack of skills among Library staff was recognized as a major obstacle in achieving this aim. The Library could not hope to provide support for users if its staff lacked the necessary expertise and confidence to do so successfully. Lack of skilled staff could also present barriers to the future development of the strategic process itself. In Phase 1, work being carried out by the E-resources Group was held up when some members of the group, who were experienced in the administration of e-resources, lacked sufficient knowledge of mobile applications to adequately carry out the testing required. Therefore the Phase 2 remit for the subgroup tasked with developing a digital media skills strategy for users was reviewed, and its new remit was to develop a programme to address this skills gap amongst Library staff.

The '23 Things' training approach

Traditional approaches to training, such as presentations and workshops, were explored by the Digital Media Skills Group. These training methods require the release of staff to participate and this can be burdensome on departments, particularly at certain times of the year. Problems in releasing staff could potentially affect the success of the programme and possibly exclude the very staff most likely to benefit from the training, i.e. staff in front-line service departments. Moreover, traditional training methods, whilst being effective in their own way, are also often passive in nature. The group felt that the likelihood of success of any programme would be greater if the content were dynamic and designed to provide opportunities for staff to collaborate and participate actively.

The '23 Things' concept was identified as a possible basis for a training programme. The 23 Things model is a self-directed online learning programme that originated at the Public Library of Charlotte and Mecklenburg County in the USA in 2006.[3] The programme was designed to encourage staff to explore Web 2.0 and new technologies in a fun way and to reward them for doing so. It is designed to be completed within a specific time. Staff were encouraged to work together and share experiences in person and through blogs. The number of 'things' to learn was set at 23, as this was considered to be wide enough in scope without overburdening participants or the course creators.

23 Things mobile

The Digital Media Skills Group was not able to identify any current use of the 23 Things concept for the exploration of mobile technologies. It therefore decided to create an in-house, self-paced 10-week online programme, to be delivered via the Library's VLE (virtual learning environment), a 23 Things blog and a practical session in the Live Lab. In terms of content, the group easily identified 23 'things' related to mobile technology. To ensure that the course was dynamic in nature, it set out to devise a range of different types of tasks and activities. An important decision taken at the outset was that the programme would cover mobile technology in general and not be specific to libraries, but where possible library initiatives would be highlighted. The emphasis would be on developing individuals' skills generally, which would provide a basis for the transfer of knowledge to the work environment.

The original list of 23 Things and the order in which they appeared changed on several occasions before the group reached the final version. There were a number of reasons for this: some new 'things' were added to the original draft list; attempts were made to ensure that the sequencing of the 'things' allowed participants to build up their skills as they progressed through the course, and to ensure that the essential skills were covered before the Live Lab sessions so that participants would get maximum benefit from the activities. The final list was broadly grouped as:

Devices:	phones/tablets, readers
Networks:	Wi-Fi, 3G, the cloud

Applications:	mobile web, apps, QR codes, e-books
Communicating:	blogs, text/instant messaging, tweets, e-mail
Sharing:	Facebook, Google+, Sharepoint
Fun stuff:	Live Lab, music, photos, videos, gaming
Other:	Library mobile strategy, reflection on things learned.

Different learning styles were accommodated and the content was made visually appealing throughout. It was delivered as a mixture of text, images and video/sound. Much of it was created by group members, but content from external sources was also used. In attempting to keep the tone and content less formal, the group employed the use of a 'Geek' character for additional, technical information and a 'Sherlock' character to identify activities that had to be undertaken.

The Live Lab is a vital component of the 23 Things training programme and it is the only compulsory element. The majority of the programme is designed to be completed on either a desktop PC or a mobile device. The Live Lab component was introduced to allow groups of participants to come together for practical experience of using mobile devices. They are asked to carry out a number of activities across a range of devices. Activities include scanning of QR codes; taking photographs and uploading them to the blog; comparing the desktop and mobile websites; and reading e-books on tablets and phones as well as on the dedicated e-readers. Those participants who do not own a mobile device are encouraged to borrow the devices, either to take home or to use in the dedicated space.

Pilot study

As the 23 Things model was the first attempt to provide training to all Library staff in an online, self-directed format, the decision was taken to test the model and its content in a small pilot study. This would provide an opportunity to identify issues with both the model and the content prior to rolling out the programme to all Library staff. A group of 23 individuals were identified for the pilot group and consisted of a cross-section of Library staff with mixed experience of mobile technology. Sessions were set up to brief the participants on the programme and to give them an understanding of what was expected of them. The pilot was an opportunity for those selected to provide honest feedback on the content, the tone and the model itself for delivering training.

The pilot began at the end of July 2012 and ran for 10 weeks. Feedback from the pilot group was very positive, with 100% of participants saying that they would recommend it to other colleagues and over 95% reporting that the Live Lab component was the most enjoyable aspect. Senior Library management have decided that this will become a mandatory training module for all staff, and it will roll out to all Library staff in 2013.

Conclusion

The Live Lab continues to be a strategic priority and it is essential that the lab is developed. Crucial to its success has been its role in engaging the wider Library staff in the strategic process. It will be central to the roll-out of the full 23 Things programme, and the MTG will look for further opportunities to promote it in order to motivate staff and allow them to extend their knowledge of new types and models of devices as the market develops. Initiatives being considered are workshops on particular aspects of mobile devices, and drop-in sessions when new purchases of devices are made or services are planned.

Beyond the Library, the Live Lab has been used as the venue for visits by colleagues from other higher and further education institutions who are interested in developing their own mobile services. The Live Lab component of the strategy and the staff training module have been of particular interest to these visitors. It appears to be a concern generally among libraries implementing mobile services to ensure that they equip their staff with the necessary skills to support these new services adequately (Kealy, 2009). The MTG will take the opportunity to collaborate with other institutions where possible and to include this aim within the strategy.

The Live Lab began as a conceptual idea in the Library's mobile strategy but, as the implementation phases have progressed, the concept has evolved. It has been developed beyond its original intent and now plays a critical part in the success of the entire mobile programme.

Notes

1 See: http://eprints.gla.ac.uk/57643/.

2 iWebKit5 is open source software for mobile webapp development: http://snippetspace.com/portfolio/iwebkit/.

3 See: http://plcmcl2-things.blogspot.com.

References

Gibson, C. and Gibbs, F. (2011) An Evaluation of Second-generation Ebook Readers, *Electronic Library*, **29** (3), 303–19.

Kealy, K. (2009) Do Library Staff Have What It Takes To Be a Librarian of the Future? *Library Management*, **30** (8/9), 572–82.

University of Glasgow Library (2010) *Mobile Strategy*, www.lib.gla.ac.uk/media/media_203782_en.pdf.

3

Mobilizing academic content online: challenges and rewards

Keren Mills and Hassan Sheikh

Introduction

The MACON (Mobilising Academic Content Online) project aimed to address the challenges involved in delivering quality academic content to mobile devices in a seamless and user-friendly manner. The project worked with EBSCO, a major content and systems provider, in order to prototype a mobile-friendly resource-discovery interface that will discover and expose quality academic content from both third-party and local collections. The project also addressed a strategic priority to present a flexible and integrated view of quality content, independent of ownership and format, to help to improve the efficiency and effectiveness of the Open University Library's systems. The MACON project was part of the JISC Mobile Infrastructure for Libraries programme.

Project context and background

The Open University (OU) is the biggest university in the UK, with more than 250,000 students. It is a distance learning university, so very few students ever visit the library building in person. Only our 1200 or so full-time research students are based on campus. Around 70% of OU students are working while studying, so their study time is tightly squeezed between work and family life. Up to 44% of our UK student population start undergraduate study without the entry qualifications they would need at a conventional university, and others start study with us after a long break from formal education. For these reasons OU Library services are focused on off-campus access to e-resources and support.

OU Library Services does not offer postal loans or document delivery to undergraduate students unless they are housebound, so easy access to e-resources is essential for them. The Helpdesk services can be contacted by phone, e-mail or webchat and are open seven days a week. The Library also provides online training and skills development.

The OU has a particular interest in mobile learner support as part of the institution-wide commitment to mobile learning. In August 2012 around 35,844 unique OU students accessed the university's StudentHome online via mobile phones, which was around 19.5% of all student visits. For OU students, the advantage of mobile access to materials is the flexibility that it offers. If you unexpectedly have a little spare time your mobile phone can help you to use it productively. If you cannot access the family computer because your children are doing their homework, your smartphone provides an alternative.

OU Library Services has had a mobile website since 2007, and was re-launched following a redesign in November 2011. It was already using EBSCO Discovery Service to provide a single search across most of the third-party and local content. The project therefore had a solid platform to start from and was able to concentrate on improving the discovery experience. EBSCO agreed to work with the Library to improve the search tool incorporated into the mobile site.

A number of publishers now offer mobile interfaces through platform-native apps or mobile-optimized websites. However, a significant barrier to the take-up of access to publishers' mobile content is that users have to first log in to the institution's system for discovery of the resource and then again to the publisher's mobile platform for resource delivery. In some cases publisher apps are IP authenticated, meaning that they will only work on the subscribing institution's campus. Prior to this project there was no easy way to bridge the gap by providing seamless transfer and access from the institutional platform to publisher's platform without forcing the users to log in multiple times. Furthermore, apps are likely to break a student's workflow within a framework of online courses.

Aims of the project

The challenge for MACON was to address the difficulties involved in delivering quality academic content to mobile devices in a seamless and user-friendly manner through a resource-discovery interface. The purpose of the project was to explore the possibilities and share the findings with the m-libraries community. The content available to OU Library users includes a wide selection of third-party material and local collections such as the Open Research Repository (ORO) and OU audiovisual materials. The local content does not include catalogue data because the majority of OU students don't have easy access to the print stock.

The project has provided the opportunity to verify what OU learners want from a single search box for mobiles and to improve the mobile search tool that is offered. Working with the EBSCO API (application program interface) has provided the Systems Development team with valuable experience which can be carried forward to improve the search/discovery experience on the main website.

As the mobile site is not yet heavily used, the Library has been able to use it to test some experimental tools, such as listing recent search terms and recently read items for individual users.

User requirements

In order to ensure that the development of the mobile discovery tool resulted in a user-friendly interface with functionality that would make it useful and attractive to OU students and staff, the project started by gathering user requirements.

Methods

Members of OU staff were invited to participate in the user requirements study via notices on the OU intranet and via a Mobile Technologies Special Interest Group (M-Tech SIG). All participants would be entered into a prize draw for one of ten Amazon vouchers. The project manager recruited some academic staff and some non-academic staff who were current OU students. The participants were asked to use an existing mobile search interface for any library searches they needed to do during January 2012. They were asked to send feedback to the project team as soon as they completed a search, telling what the experience was like and what they would want to be able to do with a mobile search tool.

In order to prompt participants to record their experiences a feedback form was provided, linked from the mobile search results page. As users had to be logged in to use the search tool, their name and the date and time were filled in for them automatically, so all they had to do was say where they undertook their search, how they found the experience and what sort of functionality they would like on the search page and results page.

Participants were asked to use their own mobile phones or small tablets where possible, but were offered an opportunity to borrow a mobile device if they did not have a suitable one.

Unfortunately, the site's device detection failed during the study, so mobile users found themselves on the desktop site rather than the mobile version of the site.

To test new developments on the search interface the Library also held a focus group. All 15 of the volunteer participants were invited, but only 2 attended.

Participants

The initial response to the invitation via the intranet and mailing list led to 16 staff volunteering to take part. One later dropped out, due to going on study leave. Of the remaining 15, only 6 participants actually provided feedback on the mobile search interface. The participants represented a range of subject areas, such as Arts, Business, Classical Studies, Earth Sciences, Educational Technology and Languages.

Results

Five participants completed the feedback form and most stated that they were satisfied with the mobile search tool as it was. One confirmed that the mobile site is used in what Google refer to as 'found time', for instance whilst queuing, travelling or waiting for someone. Another complained of being unable to sign in to the site, a problem which is likely to have been due to difficulty with entering the complex alpha numeric passwords via small touchscreen keypads. One of the aims of the MACON project is to investigate a solution for making authentication via mobile devices easier.

Participants were asked to indicate how long they had spent on the mobile search interface during a single search session. Their responses indicated that they spent either 2 to 5 minutes or 10 to 30 minutes searching. This seems to indicate that at 40% of them were willing to persist with their search, despite the limitations of small screen.

The project team also considered it important to ascertain how users might want to interact with their results once they had performed a search. It was aware that some participants were likely to be reluctant to read the full text of an article or e-book on a small screen, so the team wanted to know what they would do with their results instead. Participants were given a multiple-choice question about this in the feedback form and could select as many options as they wanted to. As shown in Figure 3.1, 44% of participants wanted to save references for later use, while only 11% wanted to read the full text on their phone.

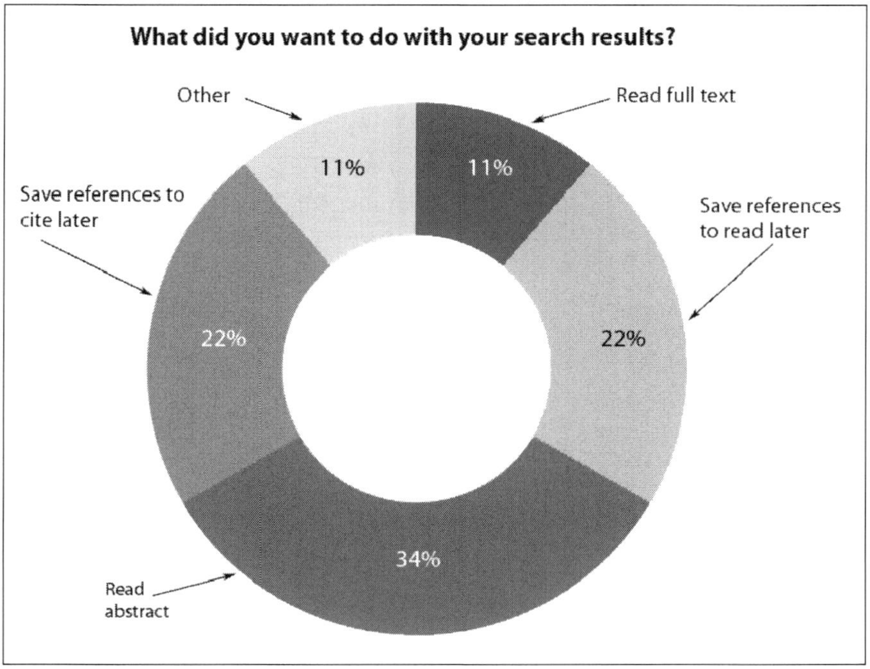

Figure 3.1 Search results preferences

The focus group participants asked for some improvements to the interface, which have since been implemented:

- Highlight search terms in the list of results (Figure 3.2).
- Expose number of search results between search box and page number navigation.
- Provide limiters as part of 'Advanced search'.
- Make it clear what is being searched (i.e. articles and e-books, not journal titles).
- Allow user to choose how many search results are shown per page (Figure 3.2).

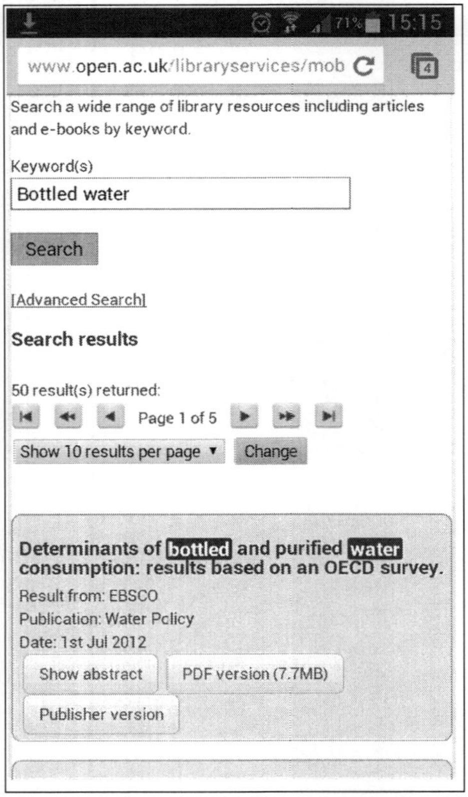

Figure 3.2 OU mobile search results interface

Project outcomes

Prototype discovery tool for delivery of academic content to mobile devices

The discovery tool draws on the EBSCO API to return search results within the OU Library Services' mobile website. The project worked with EBSCO to develop and test the prototype system using a combination of open source software (Mobile Web from MIT) and the API from EBSCO Discovery Solution (EDS). The project team sought advice from EBSCO on the use of its API and received input through an EBSCO representative on the Project Board.

The discovery tool searches across most of the Library's paid-for third-party content as well as Open Access collections and local collections. As Figure 3.2 shows, the search terms are highlighted within the results and users can read the abstract inline by tapping on the article title. Tapping 'View PDF' or 'Publisher version' takes the user to the full text of the article. Scholarly peer-reviewed articles are most commonly supplied in PDF format. The majority of smartphones, tablets and netbooks will support reading PDFs, although in most cases apps will need to be installed to enable this.

Much of the quality academic content that libraries make available to their students comes from electronic journals. The articles in these journals are most commonly provided as PDF documents, which are not ideal for reading on small screens. Often the PDFs do not allow text to reflow, which means that reading them on a mobile device can be akin to using a screen magnifier – you have a small window onto the text and you have to keep moving the text around under that window. As you cannot see the full layout of the document this can be very disorienting. When you reach the end of a paragraph it is not always clear whether you should move down or sideways. There are tools available to force existing content to reflow for mobile screens, but many licence agreements with content providers (and some copyright statements) specifically prohibit the reformatting of the materials that they supply, so libraries cannot use these tools on third-party content. The project team therefore felt that it was important to provide access to an HTML format wherever possible.

Materials in other formats can present similar difficulties, as not all devices support all common media formats. For instance, Apple's mobile devices do not support Flash, which has been a popular format for learning objects and videos.

Following user feedback, the Library plans to add check-boxes to each result, which will allow users to select results to save to a bibliographic management tool so that they can access them on other devices.

At the bottom of the screen the individual user's most recent search terms are listed so that they can easily repeat the search, along with a list of full-text items that they have accessed, so that they can easily find them again, as shown in Figure 3.3 on the next page.

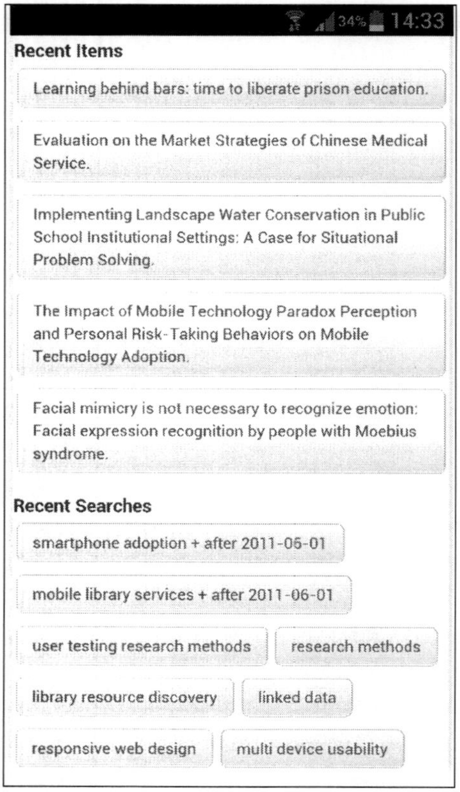

Figure 3.3 Recently viewed items and recent searches

EZproxy bookmarks to mobile databases

The mobile website now includes a list of mobile-friendly databases with persistent EZproxy links. These links direct users to authenticate through the OUSAMS authentication system and then connect them to the resource through EZproxy authentication and access software. Users can bookmark the URL and it will still work the same way.

Other authentication options were explored, as detailed on the project blog in the post titled 'Approaches to improving mobile authentication' (Mills, 2012), but the one above was the only one that complied with the Library's licence agreements and the Open University's security policies.

Good practice toolkit

A good practice toolkit for publishers and content providers delivering academic content to handheld devices has been developed as part of the project. Parts of the toolkit may also be of use to libraries. It includes good practice guidelines, basic principles and practical examples to help content providers who wish to optimize their content for delivery to mobile devices.

Technical approach

The MACON interface has been developed by adapting the Massachusetts Institute of Technology's mobile web project, which is open source software, and taking inspiration from North Carolina State University's mobile web project. The Library tweaked the code and user interface elements to meet its requirements and has also enhanced WURFL (the API) to implement three device models (and auto-detection function for mobile devices). During the project the main focus has been on refining and improving the search interface. The OU Library uses EBSCO Discovery Service (EDS) to provide a one-stop search solution to its users. The EDS API enabled integration of the search results into the OU Library mobile website, which provides efficient response for search results and avoids the confusion students can experience when they start a search on the Library website and are presented with results in a different interface. The Library has aimed to provide a simpler (but effective) search experience for its mobile users and they can expand search results fields, should they wish to see full abstracts (the dynamic expand function to view abstracts has been implemented using JQuery).

Lessons learned

Mobile discovery

This project reinforced that developing for mobile devices is complex because of the wide range of operating systems, browsers, screen resolution and processing power that are in common use at present. The Library also learned that its users do want some advanced functionality on mobile discovery tools, such as the ability to toggle between only showing search results where the full text is available and showing all results. It is always worth doing user research.

Adding audiovisual materials

Open data is not always easy to use, even when it comes from your own institution. The Library was not able to complete the inclusion of OU audiovisual materials in the search results because the metadata is only available as linked data and OU Library staff currently lack the skills to work with it.

Mobile bookmarklet

Some licence agreements with content providers or publishers prohibit storing user login credentials on users' personal devices.

Evaluation

When recruiting users to give any kind of feedback, a far higher response rate can be achieved if you can contact users directly, than if you issue the invitation to participate through a website, online noticeboard or other 'pull' media.

Good practice toolkit

The process of writing the good practice toolkit exposed the Library to different tools and methods which would have been useful during the project. The toolkit could have been of benefit to the project if it had been researched and written alongside the development and evaluation, rather than after these were completed.

Conclusions

The objective to 'address the challenges involved in delivering quality academic content to mobile devices' has been largely met. However, many of these challenges will take time to resolve, such as the issue of third-party materials being provided in formats that are not mobile friendly. This issue can be resolved only by the content providers, but many are beginning to address it.

Working on the mobile search tool has provided inspiration for ways in which the Library can improve its desktop search. For instance, we know that new learners can feel lost when they initiate a search on the Library website and their search results are presented in a different interface by the search provider. By using the EBSCO API we could bring the search results into the Library's own interface to help avoid that confusion. Working with EBSCO has been helpful because it has been able to give advice when we have had difficulty working out how to do something with its API.

Many developers now recommend designing for mobile before desktop (Gold, 2012). The OU Library experience with MACON has shown that if you consider the functionality of desktop and mobile interfaces together they can benefit from one another, and users want a consistent experience. The EBSCO representative on the project board said that it is also now designing for mobile first and then retro-fitting for desktop.

References

Gold, E. (2012) *Why Design for Mobile before Desktop?*, The Other Media, www.othermedia.com/blog/why-design-for-mobile-before-desktop,424,TA.html?category&display=TRUE&pageNo=4.

Mills, K. (2012) *Approaches to Improving Mobile Authentication*, MACON project blog, www.open.ac.uk/blogs/macon/discovery/bookmarklet/approaches-to-improving-mobile-authentication.

Websites

EBSCO API: www.ebscohost.com/discovery/api.

EBSCO Discovery Service (EDS): www.ebscohost.com/discovery.

EZproxy® authentication and access software: www.oclc.org/ezproxy.en.html.

Good practice toolkit for delivering academic content to handheld devices: www.open.ac.uk/blogs/macon/toolkit.

JISC Mobile Infrastructure for Libraries programme: www.jisc.ac.uk/whatwedo/programmes/di_informationandlibraries/emergingopportunities/Mobile%20Infrastructure%20for%20Libraries.aspx.

MIT Mobile Web: m.mit.edu/about.

North Carolina State University's Mobile Web: m.ncsu.edu.

Open Research Repository (ORO): oro.open.ac.uk.

OU audio-visual materials: http://podcast.open.ac.uk/oulearn.

OU Library Services mobile resources: www.open.ac.uk/libraryservices/m/?page=resources.

OU Library Services mobile website: www.open.ac.uk/library/m.

The Open University in facts and figures: www.open.ac.uk/about/main/the-ou-explained/facts-and-figures.

WURFL (the Wireless Universal Resource FiLe): wurfl.sourceforge.net.

4

Using iPads for a roving enquiry service: a case study on lessons learned

Rowan Williamson

Introduction

In September 2011 we purchased eight iPads. There was no specific plan for making use of them, but there were many ideas about how they could be used. As the department was working on an iPhone app it was known they would be useful for testing this, and we felt that they would have great value as tools to support our new roving enquiry service. We had not researched any other products at this time, as the iPad was the dominant tablet on the market, with little competition as yet.

This is a case study about our experiences, what we have learned and how we are developing the project. It focuses mostly on the project management side, but also includes some basic information about the technology and apps we have used.

Roving was a relatively new practice, introduced to replace staffed Helpdesks on the upper floors of the Learning Resource Centre (LRC) at our two largest campuses. In fact, roving was still in a pilot phase when we decided to introduce the iPads. The LRCs have half-hour slots throughout the day where a member of staff is timetabled to 'rove' around the LRC, the main purpose being to make the librarians available to answer enquiries on the upper floors and proactively offer help to students who are not approaching the Helpdesk on the ground floor. We worked on the basic assumption that having mobile internet access would make the roving librarians better able to answer a wider range of enquiries. Another key goal was to promote the university's new iPhone app by demonstrating to students how to download and use it, whilst answering live enquiries. We were new to the iPads and had no knowledge of their uses and full potential. A small number of staff spent some time familiarizing themselves with the devices; however, as we were not roving over the summer, it was difficult to test them in the live environment.

Implementation of the project

For this reason, our initial launch was a basic trial-and-error approach. The project plan was simple. Review the iPads from a 'security' perspective (involving set-up of PIN codes, passwords and logins), train staff in the basic navigation and use of the web browser . . . and then try it.

The first thing we did was to register a generic 'Library' iTunes account. We set up a basic 'image' for the iPads so that they all had the same content and look. This was done by using a back-up file that could be restored on a device at any time to overwrite any changes that had happened during use and maintain a consistent set-up. All the iPads were synched to this image.

We disabled the personal settings by setting a PIN code. This prevented anyone from downloading any extra content. We wanted to avoid people using the Library account to download content they wanted, or using their own accounts to download content that would be lost when we reset the iPads to the backed-up image. A PIN code with a two-minute time-out was set on each iPad. The iPad also locked whenever the cover was closed.

Basic training sessions offered information about switching on and off, navigating the desktop, opening web pages and scrolling. The training sessions were done in small groups and took about 15 minutes. We found that we were trying to fit the training in at the start of term, our busiest period, when most staff could not spare the time to attend longer sessions. Staff who had been using the iPads over the summer had discovered some useful functions which they used in the course of their own work, such as taking notes in meetings or photos on visits, but we were not sure how these could apply to a generic device being used while roving, and ran out of time to investigate further. We had all the usual apps on the desktop as it came out of the box, and some staff had added some that they were using, but these were mostly hidden from the desktop to avoid confusing those staff who had not used iPads before (the majority!). We had compiled a list of key web pages which we thought would be most useful for answering enquiries while roving, and these were bookmarked and displayed on the desktop so that people could access them quickly. These included links to pages such as the Library catalogue, the University bus timetable and the IT support pages.

Project review

In November 2011 we conducted our first review, which showed that a few key people were roving with the iPads, but the majority of staff had quickly given up taking the devices with them when roving. We needed to investigate the reasons for this. We had 'soft launched' the project and it was not compulsory to use the iPads, but we had hoped that they would be viewed as a useful tool and that staff would want to use them and, by doing so, begin to develop ideas for other ways to use them and make them more useful. Feedback showed that this was not really happening.

Lack of access to Library Management System

One of the main reasons cited for not using the iPads was the lack of access to the Library Management System and other programs needed for enquiries. Whilst websites could be useful for answering some enquiries, staff reported that students mostly found this type of information themselves, or were happy to go and search for it themselves when informed that it was available online. Beyond these types of enquiry, most questions involved IT account and Library account issues. For many of these we needed to send people down to the ground-floor Helpdesk, as the applications and programs needed to deal with such enquiries were installation based, rather than web based.

Login inconveniences

Another issue that we began to discover was that not all the e-resources worked on the iPad when accessed through the Library catalogue.

Where they did work, it could be a slow process to access them, as the devices were not set up personally and staff had to repeatedly sign in to web access, which was both fiddly and time consuming. Using a generic staff login for the Wi-Fi also limited access to resources, as the generic accounts were locked down to grant access only to basic functions, for security reasons. However, using individual logins also raised issues of account security when they were being used on a shared device which depended on the previous user remembering to log out, especially as roving shifts were only half an hour in length. Many staff found it more trouble than it was worth to log in to everything again, and either sent students to the Helpdesk for a quicker response or used one of the fixed PCs or OPACs that are on each floor of the LRCs. These had been 'Information points' used by rovers before the iPads arrived, and they were proving more popular than the iPads, since staff felt that it was quicker to take a student to the PC or OPAC than to access the information on an iPad on the spot.

Insufficient functionality

In addition, we discovered that the Library catalogue would default to the mobile version of the page, which did not allow people to sign in, and so there was little functionality available beyond searching for books. We could not show students how to check their accounts, reserve books or access e-resources. Even making notes was difficult, as the iPads were handed over at the end of a roving session and any notes on them could not easily be transferred. Sending e-mails was not possible, due to the generic login and the fact that individuals could not set up their e-mail on the devices. The iPad's biggest strength is that it can be personalized, and this was working against us.

All these factors combined to seriously limit the usefulness of the iPads. We had assumed that web access would open up a lot more possibilities for answering enquiries; however, in reality there was little we could do that students were not doing themselves on their phones and computers.

Staff inhibitions

The lack of comprehensive training and the unfamiliar technology caused many staff to feel under-confident in using the iPads, especially in front of students who, they assumed, would be far more knowledgeable than they were about the new technology. Some staff felt that their experience of roving did not match what was assumed to be happening, and so using the iPads was inappropriate for the sort of work they were doing. Staff stated that the service was misunderstood, and that the pilot had not been first properly evaluated. They pointed out that while roving they were not really providing an enquiry service, but spent more time managing behaviour, clearing litter and furniture and dealing with issues with equipment such as faulty printers on the upper floors. For these staff, an iPad was an encumbrance that got in the way of the physical element of their work.

This of course presented a 'chicken and egg' argument which was difficult to resolve, as the counter-argument was that unless we proactively promoted roving as an enquiry service and sought out students with enquiries, the students would remain unaware that we were a potential source of information and roving would remain an activity primarily focused on managing the environment. Students went to the Helpdesks for assistance, unaware that the staff roving among them were available to help, while staff were not confident that they could offer any help to those with enquiries. Seeking out and approaching students to ask them if they need help is difficult and embarrassing for many staff at the best of times. Faced with the fact that they themselves might then have to send the student to the Helpdesk anyway, most were reluctant to take such a proactive approach.

Unreliability

We discovered that our wireless network was patchy and unreliable in certain areas of the building, and its default was to log out after several minutes of inactivity. Having to constantly sign back into the Wi-Fi each time you needed to access the internet was also causing considerable delays for staff trying to answer enquiries, and left students with the impression it would be quicker to go to the Helpdesk. We also found that not all staff were taking responsibility for the basic maintenance of the iPads. They were not always plugged in to charge by the last user, and so batteries would sometimes run out during the roving session.

The result of all these issues was not only that the iPads were not being used efficiently, but that, as we had missed opportunities to get this project right, it was costing staff enthusiasm and commitment. Once gone, this was hard to win back, and so we were faced with two tough questions. Were the iPads the wrong tool for the job? Or did we just need to use them differently?

Finding solutions

We decided on reflection that many of the problems we had identified were not insurmountable. We needed to take the time to investigate the tool (iPad) and the job (roving) before we could assume that they were a bad fit.

Technically, we needed to think up some alternative solutions, and so we needed to become more familiar with how the iPads worked and how we might adapt them. We needed to come up with some basic procedures, such as signing iPads in and out, plugging in to charge at the end of sessions and daily backing up and re-imaging. And most importantly, we needed to address the staff reluctance to engage with them, and to demonstrate their usefulness again, with proper training.

iPad 'champions'

The first thing we did was to start a new group of 'iPad champions'. This was made up of staff who were more knowledgeable or enthusiastic about the devices, and whose remit would be to:

- investigate new ways of working with iPads in the department – consulting teams to identify possible uses and investigate apps and methods of working to put them into practice
- promote the use of iPads for roving and queue-busting
- manage and configure departmental iPads – charging and monitoring iPads, installing updates, adding apps, configuring the desktop and seeking advice on more technical features
- train and support staff in using the iPads.

The champions spent the next summer researching and setting up, testing and then training. The group's first recommendation was to try using a remote desktop app to allow us to remote-control a PC. This would give access to all the same programs that were available on a normal Helpdesk PC. We used the Remote Desktop Lite app, and also trialled the Pocket Cloud app for this purpose.

The 'champions' also investigated and tested a selection of useful apps, and sought ways to make better use of the existing apps and functions of the iPad. This resulted in some 'good practice guidelines' which suggested ways that staff could use the iPads' built-in functions, for example using the notes app to record enquiries or log faults that needed following up. Photos provided a quick way to make a visual note, for example of a problem or location, such as a chair that needed to be fixed or removed, or a printer error message that needed to be reported.

Wi-Fi and logins

On investigating the Wi-Fi problems, we discovered that while the local Wi-Fi was set up to time out, the Eduroam network was more stable and that, once signed in, a user would remain so. The champions set the Wi-Fi up to connect automatically to the Eduroam network under a generic login, so staff no longer needed to log in and out of it and the connection was rarely lost. It was still necessary to use generic accounts, as using personal logins was just not an option, especially for the remote desktop app, as it would be impossible (and in terms of security, risky) to set up every staff account with access to a remote desktop. Instead, the 'champions' altered the generic account so that it had access rights to key resources. We enabled the e-mail account for that login, and provided shortcuts to external login pages, so that individuals could log in to certain databases personally, rather than using the Shibboleth single sign-on option which would usually connect to the generic account by default.

Managing perceptions

With many of the technical difficulties removed, the biggest barrier to success was staff and student perceptions. The removal of the fixed PCs on the upper floors has been a big help. Staff had become accustomed to them and used them in preference to the iPads, despite the fact that the iPads were now running more efficiently. Staff became more confident as a result of the technical improvements and training, and there was no further need of fixed PCs. Staff feedback about roving suggested that the iPads were physically awkward to carry around, so we provided shoulder bags and cases with handles on the back, which staff can choose to use if they want to be 'hands free', or to make it easier to have the iPad open in your hand, ready to respond to an enquiry.

Now that we are confident in our service, more time and energy needs to go into promoting it. Branding is helping to make our rovers more recognizable to the students. We now have our Helpdesk information point logos on the iPad skins and badges on the bags, and this year we have promoted the new service in our student inductions. Students now recognize that we are there to help and this has led to an increase in the numbers of enquiries addressed to the rovers.

Lessons learned

Most of what we learned is self-evident. We all know not to rush into projects without fully investigating, we know that training is essential to the success of a project and we know that timing is everything. In the busy academic environment, where resources and time are precious, we don't always have the luxury of stepping back and doing those things that we know will help make our projects successful, but somehow we get there in the end.

There were some things we didn't know, or that we assumed we knew to be the case but were not. Every project has a learning curve and we hope that our experience may help others to avoid some pitfalls. One key thing to consider is the fact that iPads (and many other tablets) are designed to be personalized. Their usefulness and interactivity depends on this element. Libraries operating iPad loan schemes or shared resources projects like ours are therefore unable to tap in to the full potential of this technology. There are workarounds, but the leap from personal to generic is not always easy, and although the cloud is moving things further away from the traditional device, we still face the fact that many of our systems operate on an installation basis and are bound by certain security and access protocols which are inflexible. Creative thinking is needed in order to find new ways of working and to adapt the technology to suit our purpose.

Apple's catchphrase 'there's an App for that' is generally true. But it does not mean there is a 'good app for that', nor does it mean there is a 'free app for that'! Some apps, like the remote desktop, have opened up possibilities for us that we otherwise could not have taken advantage of, but there is no doubt that the desktop app is clunky, and the solution is not perfect. There may be better products out there now, but a project like ours has evolved, and has no budget. It's difficult to trial and test these things without committing to cost.

Finally, we have learned things about ourselves that we didn't know before. Despite all the problems with Wi-Fi and lack of training, it is the small things that trip you up. I strongly suspect that several of our staff were completely put off walking round with an iPad because they didn't want to carry it in our unattractive shoulder bags. So, when congratulating yourself for coming up with a handy, practical solution, always have a back-up plan just in case you have missed something obvious!

5

BYOD! We don't think so

Stephen A. Bowman

Introduction

This paper is the result of workshop sessions with the Students Union at the University of Chichester and the University of Portsmouth, examining the question 'In a fee-based environment, what will students be expecting from library and information services, learning spaces and learner support?' It looks at the question from the perspectives of 'University perceptions' and 'Student perceptions', and the consequences of these differing perceptions are discussed.

University perceptions

Since the publication of the BL/JISC report *Information Behaviour of the Researcher of the Future* (JISC, 2008), there has been a sector-wide move away from a 'Computer Labs' model of IT provision and towards ubiquitous Wi-Fi, laptops in classrooms and apps for mobile devices.

Perceived student mind-set

The report provided a snapshot of the mind-set of students in higher education at that time. This so-called 'Google Generation' (students born after 1993) is seen as 'format agnostic' (in that members just want the information that they require, in any format) and as having little interest in the 'containers' that provide the context and wrapping for information 'nuggets'. It was identified that this cohort of students has shifted decisively to digital forms of communication, preferring typing to handwriting, messaging to talking on the phone, and using its mobile devices to access information rather than the university model of desktop personal computers. The Google Generation is believed to be impatient and to have zero tolerance for delay – information and entertainment needs must be fulfilled immediately!

Personalized technology

Perceptions about the mind-set of our students have been mirrored by developments in technological access to information, in that students now have access to the web through mobile devices such as the iPhone, iPad, smartphones and tablet computers. This availability of 'personalized' mobile technologies, combined with the attributes outlined above, has led to the rise of BYOD as a possible 'solution' to perceived student information needs.

Bring Your Own Device (BYOD) is

> the policy of bringing personally owned mobile devices to [school, college, Uni] and using those devices to access [institutional] resources, as well as personal applications and data. The terms Bring Your Own Technology (BYOT) or Bring Your Own Behaviour (BYOB) are also used, because they express a broader phenomenon, which not only covers the hardware device(s), but also the software used on the device.
>
> (Wikipedia, 2012).

BYOD is also seen as the answer to technological change, in that rather than constantly trying to cope with the flood of new devices being used by students, universities should make their services and systems available across a range of platforms, 'format agnostic', so that the students can use their own mobile technologies to access the information that they require. This BYOD culture also meshes with the 'Information Commons' pedagogical model, moving away from a didactic form of instruction and towards a more collaborative paradigm, opening up a world of 'Social Learning' – rather than sitting in IT labs – and creating 'Independent Learners': the information that you want, when you want it, on your own device!

Student perceptions

Student perceptions were gleaned from a one-day workshop run in the summer of 2011 by a consortium of south-coast universities with input from their student unions (SU). The introduction of the new fees regime (whereby the funding of universities is now provided directly by the students, rather than through government-provided core allocations) was seen as having a profound effect on student expectations of library services.

Students as consumers

Presentations by the SU painted a very different technological expectation to that understood by most universities since the JISC report. The relationship between the student and the institution is now perceived as a consumer/provider interaction, rather than a partnership. Students stated: 'We are paying you £8,500 a year for our degree!' 'Why are there not 200 copies of a book if there are 200 students on the course?' 'We want people to talk to, rather than machines, we want info[rmation] given to us' (Bowman, 2012). Rather than being (or wanting to be) the 'independent learners' that universities have been encouraging, these students demand 'value for money' on their investment in a degree.

Desktops and information for all

Rather than wanting (or needing) to use their own devices to access the information needed for their studies, students stated that 'We don't want to read small screens, we don't want to have to carry laptops around, we don't want to have to look for stuff, [we] hate reading online.' The students' expectation was of a desktop machine for every student, where they want it, when they want it, along with help to find materials, as the perception was that [students] can waste time trying to find [information] themselves.

This 'consumer' orientation also flavoured students' perceptions of the kind of service that should be provided by the university, in that students could not understand why universities don't 'just tell us what we need to know'. This links in with the strong student perception that they are paying for the 'end-product' of a good degree, which should include everything that they need to acquire this while they are at university:

> a first class library service, free printing, and real books on the shelves, all course materials (stationery etc.) supplied for every student, quiet zones for when we want to be quiet and noisy areas for when we want to be noisy, along with 24 hour opening – 'Everywhere else does it.' (Bowman, 2012)

The JISC report (JISC, 2008) noted the perception that

> young people have unsophisticated mental maps of what the internet is, often failing to appreciate that it is a collection of networked resources from different providers, as a result, the search engine, be that Yahoo or Google, becomes the primary brand that they associate with the internet.

Students in the workshops reinforced this perception by demanding 'access to all of the info without having to move. No more than 3 clicks away – accessibility [seen as] key for all online information'. Student perceptions of this consumer relationship centred on having 'as many retakes as are needed to pass, and our degree (that we have paid for) at the end'.

Conclusion

Universities have been moving away from 'computer lab' provision of information technology access since the 2008 JISC report. With the increased availability of personal mobile devices and the increased costs of providing IT equipment, BYOD is seen as a viable way forward for future 'student-centred' provision of IT services.

However, students' perceptions and expectations as expressed through the workshops run during 2011 have posited a very different 'student-centred' IT landscape. The new fees regime (introduced in 2012) has changed the student/university relationship from one of mutual partnership to one of consumer/producer. Students do not want to use their own devices to access information. They expect their universities to provide access through institutional IT equipment that they, the students, have 'paid for' through their fees. There is also an expectation that all library and IT services that they will use while at university should be provided free of charge.

There is a strong perception on the part of students that they should be 'given' the degrees that they have 'paid for', rather than have to become the 'independent learners' that universities have been encouraging. The new fees regime does not provide any additional income for the universities, but has vastly raised the expectations of students, who see it as 'a right to use whatever they want, whenever they want it (at no extra cost)'. The perception (and strong assertion) conveyed by the SUs was that students will require everything to be provided for them because they have paid for it.

In essence, students will want everything they have always wanted ... AND MORE!

Bibliography

Bowman, S. A. (2012) *BYOD! We Don't Think So*, M-Libraries 4 Conference, Milton Keynes, www.m-libraries.org/bowman.

JISC (2008) *Information Behaviour of the Researcher of the Future*, CIBER, www.jisc.ac.uk/media/documents/programmes/reppres/ggworkpackageii.pdf.

Wikipedia (2012) *BYOD Definition*, http://en.wikipedia.org/wiki/Bring_your_own_device.

6

Bridge over troubled waters: QR coding the collection for student satisfaction

Neil Ford

Introduction

Since 2012 Bournemouth University (BU) Library has been highlighting available, relevant e-books by embedding QR codes in its printed book collection. This chapter will focus on the drivers and intended outcomes of the project and will discuss the characteristics of successful QR campaigns that we have considered in order to achieve our aims. A more practical discussion of project implementation is in press at the time of writing (Ford, 2013).

Bridging the gap between traditional skills and digital literacy

In an increasingly digital information landscape, we believe that digital literacy and student satisfaction are closely linked. Over the past decade, BU has invested heavily in e-resources which has vastly increased the consumption of library information. Annual e-book views have increased exponentially, to over 2.5 million; however, feedback suggests that not all students use e-resources to support their learning.

We attribute this, at least in part, to the information-seeking behaviour and digital literacy of some students. We encourage students to search our online discovery tools for the *full* range of print and electronic resources, but techniques like visiting the Library and browsing the shelves are still prevalent (Korobili, Malliari and

Zapounidou, 2011). Despite a collection of over 100,000 e-books, the limited availability of electronic textbooks continues to drive visits to the print collection. The discovery of print books through shelf browsing depends on how the library has classified the books. Classification is highly subjective (Keshet, 2011) and materials are not always grouped according to the concepts that students relate to.

Comments like 'buy more copies of up-to-date books!' for subjects that are well covered by our e-book collections suggest that some students assess library coverage of their subject by what they see on the shelves. We were therefore keen to encourage those students who were not finding physical books to develop the resilience to explore digital alternatives.

QR codes bridge the physical/virtual divide

Towards the end of 2011 we explored QR codes for this purpose. QR codes are two-dimensional graphical codes that encode a string of characters. Commonly, this string is a URL that takes users to a web page in the browser of the device that is used to scan the code. The proliferation of mobile internet devices has led to the increasing use of QR codes in marketing (Tolliver-Walker, 2011), where they are thought to have potential as a 'bridge between offline and mobile media' (Okazaki, Li and Hirose, 2012, p. 102). QR codes are often misused in marketing campaigns (Charlton, 2013; Eden, 2013). Following attendance at a JISC-funded workshop on QR codes (Knox, 2013), we made efforts to ensure that our project met the conditions required for a successful QR code campaign.

User group

Firstly, we asked whether we had a user group that would use QR codes. Educause recently estimated that two-thirds of college students owned a smartphone (2011) and later highlighted increasing numbers of students owning smartphones and tablets and using them for academic purposes (Dahlstrom et al., 2012). We are thought to be entering a 'post-modality' era, where technology is challenging traditional notions of learning environments, enabling an 'anytime, anywhere' approach to learning (Dahlstrom et al., 2012, p. 7). Research at BU estimated that 75% of its students own a smartphone (Tedd, Wardle and Heppell, 2012). More generally, the use of QR codes is predicted to mature into wider adoption over the coming years (Gartner, 2011). As technology infiltrates into the lives of our learners it may offer us opportunities to construct learning by using devices, media and skills that our students already have (Beetham, McGill and Littlejohn, 2009; Trinder et al., 2008).

Figure 6.1
Bournemouth University QR code sample

Mobile content

Failing to provide a *mobile-enabled* target is a common fault of QR campaigns. Users are frustrated by web pages that are not designed to be viewed on mobile devices (Edwards, 2011). In the light of the varying usability of academic e-book platforms on mobile devices (Lam et al., 2009), we wanted our QR codes to direct students to useful and usable mobile content. Fortunately, this coincided with the launch of a mobile interface to our online catalogue (Capita Prism) (Westwood, 2011). The mobile interface (see Figure 6.1) displays catalogue search results optimized for mobile browsers, showing students what e-resources are available whilst they are in the stacks. It also gives them a list of resources that they can view later from devices better suited to reading.

Incentive and user motivation

Only a small number of QR campaigns include any incentive for users to scan the code (Okazaki et al., 2012, p. 106). We wanted to capitalize on BU students' motivation to find books relevant to their units of study, so we grounded our project in unit reading lists. Analysing materials on reading lists determined the subject area of the unit. We then constructed searches for e-books relevant to the subjects. Given that students can struggle with matching subject headings to search terms (Bates, 1977), we believed that recognizing the name of their unit would be an important factor in motivation to scan the code. We therefore followed the nomenclature of units of study on our labels. Similarly, we wanted students to see QR codes in the areas where they would most likely be searching for books. We used the shelfmark of the core textbook listed on the unit reading list to locate the QR code.

Code design and usability (size, data density and readability)

QR code design can have a significant effect on how easily they can be scanned. We used black and white images to ensure high contrast for minimum processing time (Liu, Yang and Liu, 2008). Producing the code graphics as large as possible and surrounding them with 'white space' makes it easier for the device to interpret the code (Ward, 2011) and easier for users to locate the code in their phone's viewfinder. The density (or number of pixels) of the code also affects how easy it is to scan (Kindberg, 2007). We used URL shortening to reduce the number of characters in our catalogue search URLs and, thus, the code density. This also gave us a shortened URL that could be printed on our labels for those without smartphones.

As well as the technical aspects of usability, we also took measures to ensure that our labels were simple, clear and accessible. For example, we implemented changes to our proposed colour and font following consultation about accessibility with our Additional Learning Support department.

Environment

Environmental conditions like lighting and internet access can affect the user experience (Liu et al., 2008). Although most devices can now store the results of a scan for later use (Eden, 2013), we felt it was important for students to be able to access available e-books whilst at the shelves. Fortunately, Wi-Fi access in BU libraries was recently enhanced by a significant programme of IT infrastructure improvement, so we were confident that there would be a good user experience over Wi-Fi. The mobile interface to which we were directing users employs caching of data to optimize search speeds in low bandwidth environments, such as 3G (Westwood, 2011). In order to test the user experience, we created prototype QR codes and scanned them using a range of devices and from various positions in the library.

Conclusions

It is too early to fully evaluate the impact of our project on student perceptions of the availability of material; however, we can tell that the codes are regularly being scanned and have also received positive comments from students and academics. We conclude that with an appropriate user group, 'mobile-friendly' content, user motivation and consideration for design and the library environment, QR codes offer potential in signposting library users to e-resources as we transition from printed to increasingly digital media.

References

Bates, M. J. (1977) Factors Affecting Subject Catalog Search Success, *Journal of the American Society for Information Science*, **28** (3), 161–9.

Beetham, H., McGill, L. and Littlejohn, A. (2009) *Thriving in the 21st Century: Learning Literacies for the Digital Age (LLiDA project)*, Caledonian Academy, Glasgow Caledonian University.

Charlton, G. (2013) *10 Examples of QR Code Madness*, http://econsultancy.com/blog/10818-10-examples-of-qr-code-madness?utm_campaign=bloglikes&utm_medium=socialnetwork&utm_source=facebook.

Dahlstrom, E., de Boor, T., Grunwald, P. and Vockley, M. (2012) *ECAR Study of Undergraduate Students and Information Technology*, Educause.

Eden, T. (2013,) Top 10 Myths about QR Codes, *Net Magazine*, 22 October, www.netmagazine.com/features/top-10-myths-about-qr-codes#null.

Educause (2011) *7 Things You Should Know about Mobile App Development*, www.educause.edu/library/resources/7-things-you-should-know-about-mobile-app-development.

Edwards, C. (2011) *Make a Friendly Mobile Website for Your QR Code Landing Page*, www.708media.com/qrcode/qr-code-landing-page-mobile-friendly/.

Ford, N. (2013) Embedding QR Codes in the Bournemouth University Print Collection, *SCONUL Focus* 56, www.sconul.ac.uk/publications.

Gartner (2011) *Gartner's 2011 Hype Cycle Special Report Evaluates the Maturity of 1,900 Technologies*, www.gartner.com/it/page.jsp?id=1763814.

Keshet, Y. (2011) Classification Systems in the Light of Sociology of Knowledge. *Journal of Documentation*, 67 (1), 144–58.

Kindberg, T. (2007) *Mobile Codes: standards and guidelines a discussion document*, www.mobilecodes.org/StandardsDiscussion.pdf.

Knox, S. J. (2013) Using QR Codes in Higher Education Workshop – 31 Jan 2012, http://blogs.bournemouth.ac.uk/business/2011/12/13/using-qr-codes-in-higher-education-workshop-31-jan-2012/.

Korobili, S., Malliari, A. and Zapounidou, S. (2011) Factors That Influence Information-seeking Behavior: the case of Greek graduate students, *Journal of Academic Librarianship*, 37 (2), 155–65.

Lam, P., Lam, S. L., Lam, J. and McNaught, C. (2009) Usability and Usefulness of eBooks on PPCs: how students' opinions vary over time, *Australasian Journal of Educational Technology*, 25 (1), 30–44.

Liu, Y., Yang, J. and Liu, M. (2008) *Recognition of QR Code with Mobile Phones*, IEEE, 204–5.

Okazaki, S., Li, H. and Hirose, M. (2012) Benchmarking the Use of QR Code in Mobile Promotion, *Journal of Advertising Research*, 52 (1), 102–17. doi: 10.2501/JAR-52-1-1021117.

Tedd, A., Wardle, J. and Heppell, S. (2012) *A Step Change in Bournemouth University's Use of Technology to Enhance Learning*, Bournemouth University.

Tolliver-Walker, H. (2011) Making Best Use of QR Codes: gleaning lessons from the latest data, *Seybold Report: Analyzing Publishing Technologies*, 11 (23), 2–7.

Trinder, K., Guiller, J., Margaryan, A., Littlejohn, A. and Nicol, D. (2008) Learning from Digital Natives: bridging formal and informal learning, *Higher Education Academy Research project, Final report – 12 May 2008*.

Ward, M. (2011) 4 Quick Tips on Implementing QR Codes, *Target Marketing*, 34 (10), 11.

Westwood, R. (2011) *Going Live with the New Prism Mobile Interface ...*, http://blogs.capita-libraries.co.uk/prism/2011/10/07/going-live-with-the-new-prism-mobile-interface-2/.

7

If you tweet will they follow? Promoting library resources and services to a mobile audience through social media

David Honeybone

Introduction

This chapter details the implementation and operation of the @unilibrary Twitter account for the University of Melbourne Library. In 2010 the account was established, and initially linked to the Library's Facebook account. In June 2011 this changed, with a librarian being introduced to actively monitor the account and interact with patrons.

The following discussion is a personal account of how the account developed: replying to complaints; harnessing the full potential of Twitter through imaginative and informative tweets (as opposed to institutional broadcasting); developing a distinct 'voice'; actively searching for and interacting with new and existing library users; chasing the elusive retweet; what resources should be highlighted. And, ultimately, how this impacts on an already time-poor librarian.

What is Twitter?

Twitter was launched in 2006 and is a free, micro-blogging social media tool available across a myriad of mobile devices including iPhone, iPad, Android, BlackBerry, Windows 7 and SMS. Through the brevity of its messages or tweets (140-character limit), it engages naturally with a mobile audience and acts as a powerful networking tool through its speed and reach.

In Australia, 20 million people (89%) own a mobile phone (ACMA,

2011) with 2.1 million (1%) being active Twitter users (Cowling, 2012) and 11.8 million on Facebook (Cowling, 2012).

The University of Melbourne and social media

The University of Melbourne is one of the oldest universities in Australia and home to 50,000 students and 7000 staff. There are ten library locations.

In August 2010 the Library marketing manager established Facebook and Twitter accounts for the Library. However, due to time constraints the Twitter account was simply linked to the Facebook account, so anything posted to Facebook also appeared on Twitter.

The adoption of social media was undertaken so that the Library could:

- interact with customers who felt more comfortable using social media as a means of accessing information
- appear more human and accessible
- be where its customers are
- treat face-to-face and online queries the same, and hence provide equity of access.

This adoption also coincided with the introduction of Library Chat (real-time online conversations) and physical changes in the main Arts library, with the development of a flexible open space (Ellis and Kealy, 2011).

My interest in Twitter arose from the social media module undertaken as part of a Masters in Information Management. The main assignment for this subject centred on a case study reviewing the use of social media at three university libraries (University of Melbourne, University of Melbourne Law Library and Warwick University in the UK).

My research led me to conclude that Twitter linked to Facebook is a waste of a social media medium. I submitted a proposal to the marketing manager outlining the fact that Twitter's popularity is based on interaction, personal asides and conversation between users. It is therefore better to generate conversations, pose questions, share links, retweet posts and reply to people. Duplicating posts from Facebook creates no added value and hence little incentive for people to follow a Twitter account.

My proposal was accepted and I became responsible for the Twitter account in early June 2011.

In the beginning

Lines of communication were established with Library staff through e-mail and Yammer, drawing their attention to this new communication vehicle and requesting items for promotion to be forwarded.

The University of Melbourne's Social Media Policy (MARCOMS, 2010) proved very useful in providing guidelines regarding dos and don'ts. The recommendation to do only good made a particular impression.

To build followers, it was important to follow relevant large Twitter accounts such as the State Library of Victoria @Library_Vic and the National Library of Australia @nlagovau, as well as the main University account. Interaction with accounts with a large number of followers is a constructive way to gather new followers. There are also any number of social media bloggers with ideas on how to blog effectively.

Tools for the job

The library already had a bit.ly account (URL shortener) and this also allowed for statistical analysis of link hits. In order to source items of interest I used paper.li, which aggregates items in an easy-to-read digital newspaper format, from my own Twitter account. I also relied on blog feeds to source relevant articles. In late 2012 I also used Tweetdeck and Buffer to monitor tweets and time-release them, respectively.

Voice

The voice of a Twitter account is important. A distinctive tone and attitude distinguishes an account and provides a character of its own. I didn't really give too much thought to this and really tweeted as myself, with a combination of humour and information. With only 140 characters to express yourself, it is important to make an impact. A colleague came to describe it as the 'edgy voice' of the Library. The success was based on providing information and engaging with followers in a conversational, chatty manner. The success of the account had much to do with the juxtaposition of this edgy, funny voice, set against the perceived notion of the Library as a rather stuffy, traditional institution (Figures 7.1 and 7.2 on the next page).

Figure 7.1 Tweet from @unilibrary

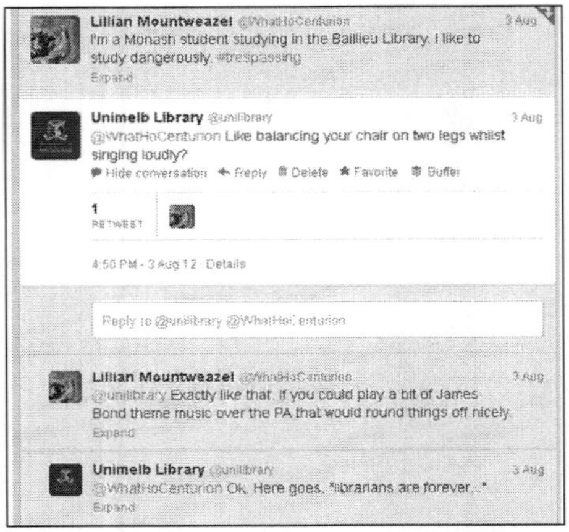

Figure 7.2 Twitter conversation between @unilibrary and a user

Searching and interaction

Searches were set up to locate any mention of the University libraries or related items (Figure 7.3). Twitter searches ran three times a day: morning, lunch and late afternoon. Due to the names of the libraries, various permutations had to be introduced to ensure that nothing was missed. Searching also helped to retrieve any complaints and enabled them to be addressed promptly (Figure 7.4).

Overall, the strategy to attract followers and, for want of a better word, protect the brand, worked. I would deliberately seek out any conversation concerning the University libraries and interact with people irrespective of whether or not it was a criticism. The reaction, in the main, was positive and initially one of surprise. Nobody expected the Library to have a Twitter account, least of all one that talked to them on a personal level.

What to tweet?

Content of tweets could be divided into three fairly broad categories: Library-related news, general information and humour. Library-specific content included the following information: Libguides, classes, search tips, new databases, database trials, new facilities, events and apps. In particular, any information regarding mobile technology proved very popular.

Reach

The reach of Twitter is probably what distinguishes it from any other form of social media. The ability to spread a message through multiple

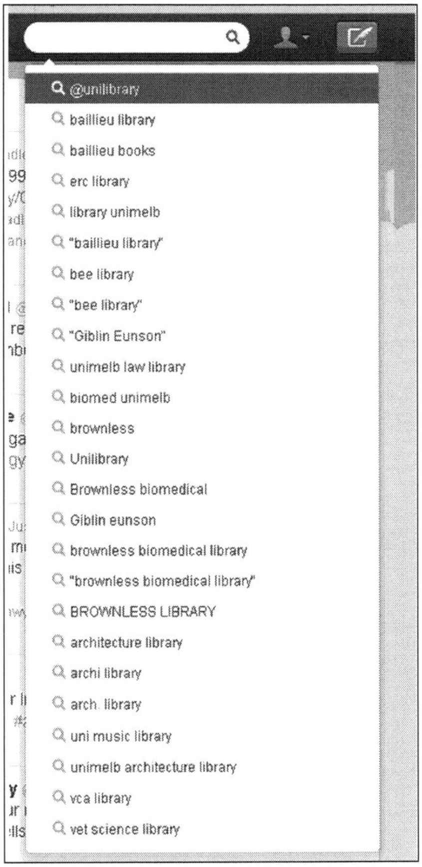

Figure 7.3 Twitter search list for University of Melbourne libraries

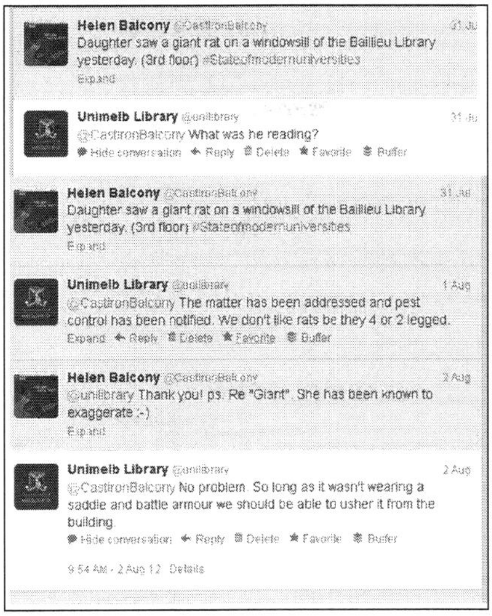

Figure 7.4 Addressing a complaint from one of the libraries

Figure 7.5 Tweeting the Melbourne earthquake

retweets harnesses the full capacity of Twitter. A viral tweet, whether planned or accidental, raises the profile of a Twitter account.

At about 9 p.m. on 19 June 2012, Melbourne experienced an earth tremor of some force. A colleague of mine who has access to the Library account tweeted about it (Figure 7.5). The response was phenomenal, eventually resulting in more than 650 retweets and more than 100 new followers in the space of 24 hours (Figure 7.6 on the next page)!

Time constraints

Responsibility for the Twitter account was undertaken voluntarily. It was never part of my position description. But, as the account became more

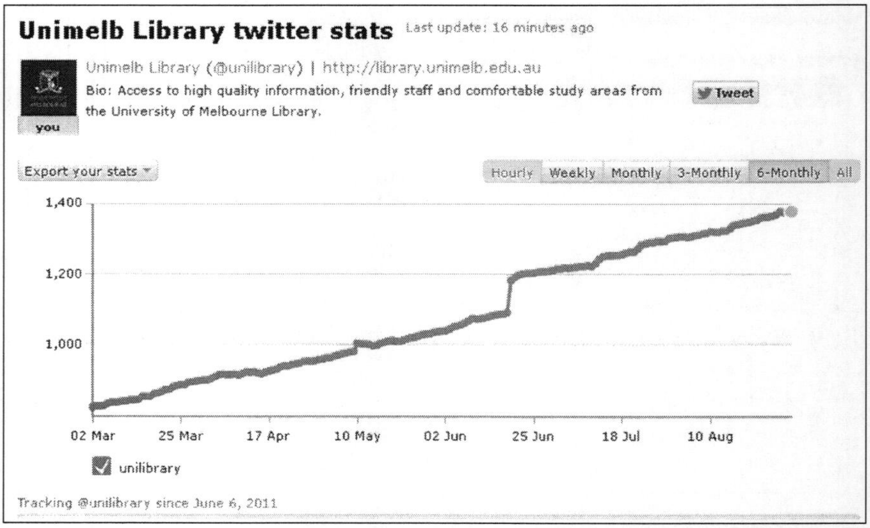

Figure 7.6 Leap in @unilibrary Twitter stats following earthquake tweet (see Figure 7.5)

successful it became an increasing impost, time-wise. It initially took up an hour, spread over the working day, and I also chose to monitor the account out of office hours. Ultimately I became a victim of my own success and I have now, regretfully, stepped away from the account altogether. Although enjoyable, it was becoming unsustainable for a liaison librarian with responsibility for four large subject portfolios. The case for a social media librarian (Choi, 2012) is well made but few, if any, libraries appear to have the funds or inclination to support such a position.

A Library Twitter Team is now in place on a weekly roster. Initial concerns included not enough time to tweet effectively and constantly, and what to tweet. There was also a concern about voice being lost. This was hard to gauge but there appears to have been little or no negative feedback and follower numbers continue to increase. Plans are also in place to train interested members of staff to become part of the Twitter pool.

Conclusion

If you tweet, will they follow? Ultimately, yes. Academic libraries are increasingly faced with a mobile audience who do not have to physically access a library in order to utilize resources. The adoption of social media effectively promotes library resources and services by supporting the traditionally static library website and creates immediate reciprocal communication to this mobile audience.

I took an idea and ran with it: I had no real strategy to start with and it worked. I was fortunate to have a lot of freedom and the account, by library standards, became very successful, gained national publicity (ABC, 2012) and resulted in a broader positive shift in attitude towards the Library. Within the space of 12 months 1000+ followers were gained, and Library users are following and using linked resources.

References

ABC (2012) RN Drive Twitterati, Sydney: ABC Radio, June 8.

ACMA (2011) The Emerging Mobile Telecommunications Service Market in Australia: executive summary, *Communications Report 2010–11 Series, Report 3*, Canberra.

Choi, C. (2012) Is Your Library Ready for a Social Media Librarian? *ALIA Biennial 2012*, Sydney.

Cowling, D. (2012) *Social Media Statistics Australia – December 2012*, www.socialmedianews.com.au.

Ellis, J. and Kealy, K. (2011) Building a 'Storehouse of Wisdom' with an Environmental Conscience: a work in progress, *World Library and Information Congress: 77th IFLA General Congress and Assembly, Puerto Rico*.

MARCOMS (2010) *University of Melbourne Social Media Guidelines*, http://socialmedia.unimelb.edu.au/wp-content/uploads/2010/08/UoM-Social-Media-Guidelines.pdf.

8

Transforming the service: supporting mobile devices with minimal budget and time

Georgina Parsons

Introduction

Brunel University is a well-established institution with library services that have received increasingly high scores in the UK's National Student Survey, an annual survey of final-year undergraduates across the UK's publicly funded higher education establishments, the results of which are available to universities and prospective students. Eighty-one per cent agreement with the statement 'The library resources and services are good enough for my needs' in 2009 has increased to 91% in 2012, following the introduction of 24-hour term-time opening, a significant refurbishment, the implementation of RFID (radio frequency identification) for self-issue of books and automated handling and sorting of returns and the introduction of an eating area within the Library. While most areas of the service have been greatly improved in recent years, the focus on developing mobile support continues, but there has been little budget for it. In this area, the Library has maintained a focus on smaller projects, using various low-cost and low-intensity (in terms of staff time) means to enhance its service.

Roving with iPads

One side of the mobile focus was on staff services, and we approached this by buying two iPads in 2011 for use by roving Library staff. They were set up by the Library Systems Team so that account information was securely held centrally and are kept with the Customer Services supervisors. Staff are scheduled to rove in pairs for one-hour slots throughout busy periods of each day, and they began by taking one iPad with them on each rove.

Initially there was some resistance to this amongst those staff who were unfamiliar with tablets, although basic training was given by the supervisors. The most problematic area was the institutional Wi-Fi set-up, whereby each device had to be checked out to the individual who was roving so that they could log in to the Wi-Fi network with their own network account; it could take up to five minutes for the checkout to take effect on the iPad (with no alert when it was ready), leading to frustration amongst those who expected it to work instantly, and inevitable instances where the devices were left logged in under a username at the end of the rove.

The iPads were set up with four main shortcuts on the home screen:

1 The Connect portal: the institutional portal for registering mobile devices on the University network. This was used by staff to check out the iPad, and also to demonstrate the portal to users, as Wi-Fi connections are a frequent difficulty amongst students.
2 The Library catalogue: this is used for print item searching, floor plans and item location and answering account queries.
3 The Library website: this provides opening hours, electronic resource searching through Summon, contact information, workshop information, etc.
4 LibAnswers and Reference Analytics: this pair of mobile-friendly products serves as both our online knowledge base, so that staff are ready with the answers to common questions, and our enquiry-logging software, so that staff can record enquiry statistics on the go, with just a few taps of the screen.

TeamViewer was also initially installed for those members of staff who wanted more information than is found in the Library catalogue; they were shown how to connect remotely to their PC so as to use the Library Management System's staff client (or any other software they used on their PC). Staff quickly stopped using this, as the relatively small touchscreen interface with remote mouse usage was impractical.

The iPads can also be used to demonstrate our mobile offerings to students, i.e. showing them what is available online and raising awareness of features such as the floor plans, QR codes in the catalogue and other mobile offerings mentioned in this chapter.

Library rovers recorded 355 enquiries using the iPads in the first (2011–12) academic year.

Filling the mobile website gap

The other side of the mobile coin concerns students' ability to access Library resources themselves on mobile devices. The Library does not have a mobile-friendly website, due to the content management system used by the University and the branding guidelines with which the Library must comply (an extension into the mobile-friendly realm is to be a future project). The development of a separate mobile website was beyond the scope of the Library's resources, there being no developer to create one, nor money to pay for external development. However, we explored many free or cheap routes for making the Library's services accessible to users on the go and via mobile devices.

Facebook

Students use Facebook from their mobile devices, and the Library already had a Facebook page. We ensured that this was promoted on the Library's web pages and were also able to purchase (on an early-bird deal) the SirsiDynix Social Library app to integrate the catalogue holdings with the Facebook page. Unfortunately, page apps are not currently viewable on mobile devices, due to Facebook's set-up, so this did not turn out to be a mobile move!

Twitter

Another popular mobile site is Twitter, with Twitter clients available on most smartphones. Brunel Library has been tweeting since 2009 and has built up a good following (1225 followers as of the start of December 2012), with conversational but informative style and quick responses. Several members of staff monitor the account and it does not take up much time. We also search for tweets about (but not to) the Library, although we intervene only when we strongly feel this is necessary.

BookMyne

Whilst the Library catalogue has improved in recent years, it is not fully mobile friendly. The supplier, SirsiDynix, has released a free mobile app, BookMyne, for catalogue and account functions (for iOS in 2010 and Android in 2012). The Library installed this in order to offer its users an alternative means of access. Due to certain display issues in both apps, and performance issues on the Android app, it has not been heavily promoted and there are few users; however, it provides an example of the free offerings that may be available and potentially very useful after some further development.

LibraryElf

An institutional subscription to LibraryElf is very cheap and offers users another way to manage their accounts. They can sign up for additional e-mail or RSS notices (or SMS, as of December 2012 at Brunel University Library) for hold, overdue or pre-overdue notifications. There is also a mobile site from LibraryElf which can be bookmarked for quick access to a calendar showing item due dates and a link through to renewals.

Summon

Brunel Library chose Summon as its resource discovery tool in 2010. This automatically redirects to a mobile-friendly interface when users access it from their device. This is another tool that we can promote for searching from a mobile device and covers e-resources as well as print holdings.

Quick response (QR) codes

We feel that QR codes can be beneficial when used as a quick alternative means of accessing information (they must not be the only means), and they have been introduced gradually. We observed that many users were writing long shelfmarks on their hands or notepads, or typing them into their phones, so QR codes were added to the catalogue for easy transfer of the item title and shelfmark into a text file on their devices. We received positive feedback on this. We are also experimenting with QR codes on the shelves, next to popular reading-list items, either for direct links to place a hold if no copies remain, or to link through to the e-book version of a title where this option exists.

LibAnswers

As mentioned above, this is our virtual reference platform, where users can search Library FAQs and submit a question if they are unable to find the information they need. It has a very simple mobile interface, so we can point users towards this as the easiest mobile access point for library information. We also offer SMS support through LibAnswers, and this includes an instant messaging/chat function which works very well on mobile devices for support on the go; this has proved very popular within the Library when users are unable to find books, access rooms, wish to report noise/environmental issues, etc.

Subscribed content

The Library website also contains a page advising users of subscribed-to e-resources that have mobile platforms or mobile-friendly websites. The information is not exhaustive but can provide students with helpful advice on easier ways to access resources if they wish from their mobile devices, e.g. when commuting.

Conclusion

Brunel Library has achieved a relatively wide breadth of mobile offerings which demonstrate to its users that it is stepping into the mobile domain and offering flexible services that the Library is constantly reviewing, developing and improving. Whilst the offerings may not be complete, specifically with the lack of a true mobile website, we are showing our keenness to develop mobile services and have been able to offer a large suite of service improvements in the mobile arena, using a small budget and little staff time.

Website references

Brunel Library catalogue: www.brunel.ac.uk/library_catalogue.

Brunel Library mobile-friendly e-resources,
www.brunel.ac.uk/services/library/ databases/mobile.
LibAnswers at Brunel Library, http://libanswers.brunel.ac.uk.
The National Student Survey: information and results,
www. thestudentsurvey.com/ faqs/faqs_1.html and http://unistats.direct.gov.uk/.

Part 2

Inspiration

9

M-education reaching the unreached: a Government of India initiative

Parveen Babbar and Seema Chandhok

Introduction

Mobile devices are playing a significant role in delivering education in India through the use of mobile and wireless technologies. The importance of this medium is steadily being realized by the Government of India, which is working towards mobile education (m-education or m-learning). The government is encouraging m-education by means of distributed peer collaboration over wireless devices, tablets and desktop computers, so as to create opportunities for discovery and education.

The government is working to meet learners' needs to acquire knowledge and skills for life in a complex and globalized world. Steps are being taken to create active m-learning communities and to target models and solutions for mass training in the use of m-education. Various plans are afoot to transform schools, universities and other educational institutions using interactive, personalized and distributed learning resources delivered via mobile devices.

Following the UNESCO plan of One Laptop Per Child (OLPC), the government has launched a low-cost tablet computing device. The device has been developed as part of the National Mission on Education through Information and Communication Technology (NME-ICT), which aims to link 25,000 colleges and 400 universities on the subcontinent in an e-learning programme via the Sakshat portal. The tablet is known as Aakash (Sanskrit for 'sky'). The aim in launching this cheap tablet, which is cheaper than other tablets and affordable for many millions of Indians, is to boost education and make rural students familiar with new technology. The Human Resource Development Ministry has floated a Cabinet note to get approval for the procurement of 5 million Aakash tablets for distribution among college students by the end of the year 2012, at a cost of over US$20 million. The commercial version of the tablet will be retailed under the brand name UbiSlate. With the launch of this tablet PC, Indian schools and colleges will be able to discover and leverage global educational content.

This chapter looks at the innovative use of wireless and mobile technologies in education. The launch of the Aakash tablet will be explored, and its spread and significance in contributing to m-education in India will be highlighted.

A mobile world

We all know that the global use of mobiles phones has increased. The number of mobile phone subscriptions worldwide, both pre-paid and post-paid, has grown from 1 billion in the year 2000 to over 6 billion in 2012.[1] Of these 6 billion subscriptions, nearly 5 billion are in developing countries. Ownership of multiple subscriptions is becoming increasingly common, suggesting that the total number of subscriptions will soon exceed that of the global human population (7 billion). Concurrently, there are 1.8 billion television sets, 1.3 billion PCs and 1.1 billion landline phones, making a combined global total of 4.2 billion of these devices. But even this is far fewer than the 4.8 billion mobile handsets presently in use worldwide.[2]

With this growth, mobile learning space is also rapidly growing in popularity in India and thus playing a significant role in the delivery of education. The significance of the mobile medium is slowly and steadily being realized by players in a variety of industries, who are constantly developing applications that are needed to work towards mobile education. The m-learning landscape is shaped by the availability of technologies that support flexible, accessible, personalized education. M-learning makes good use of the characteristics of mobile devices – including ubiquity, access, richness, efficiency, flexibility, reliability and interactivity – to reach learners. One of the major m-learning initiatives is government support for the distribution of low-cost tablets to students.

Anticipating its future potential, several telecoms companies have now started offering m-education services such as English lessons, dial-in tutorials, school syllabi, question sets, vocabulary tutorials, exam tips, exam result alerts and education for the disabled. Studies show that penetration of m-learning into Mobile Value Added Services (MVAS) is around 7–12% in India, and is expected to grow significantly in the coming years.[3]

Indian telecoms industry

India is one of the largest telecoms markets in the world, with more than 18 million new subscribers every month, and the telecoms sector has continued to emerge as the primary engine of economic growth, contributing to nearly 3% of Indian GDP. The sector has undergone a major transformation through significant policy reforms, particularly under the National Telecom Policy adopted in 1999. The National Telecom Policy – 2012 is also designed to ensure that India plays an effective role and transforms the socio-economic sphere, accelerating equitable and inclusive economic growth by placing special emphasis on the provision of affordable and quality telecommunication services in rural and remote areas. Growth has been explosive in the wireless segment, with teledensity levels reaching 75.42% and the subscriber base reaching more than 913 million. India is the world's third-largest internet user base, with over 137 million subscribers. The telecoms industry aims to increase rural teledensity from the current level of around 39% to 70% by the year 2017, and 100% by the year 2020. It also aims to provide affordable and reliable broadband-on-demand by 2015 and to achieve 175 million broadband connections by 2017 and 600 million by 2020 at minimum download speed of 2 Mbps, making higher speeds of at least 100 Mbps available on demand.[4]

According to a 2012 report from the World Bank, India has 70 mobile phone subscriptions per 100 people, of which 92% are pre-paid connections. Mobile phone networks currently cover 83% of Indians, although obviously this number still lags far behind other countries.[5]

A conducive business environment, favourable demographic outlook and political stability have all contributed to the phenomenal growth of the Indian telecoms industry. India also has the distinction of having the world's lowest call rates, at 1 US cent; the fastest sale of mobile phones per week; the world's cheapest mobile handset, at US$10; and the world's most affordable colour phone, at US$17.

Indian telecoms: future predictions for the industry

The total investment in broadband roll-out is expected to be US$20 billion, with another US$20 billion to be invested in augmenting the transmission network. India's current MVAS industry has an estimated size of US$2.7 billion. It derives 80% of its revenues from top products like game applications, music downloads, etc. The MVAS industry is estimated to grow to US$10.8 billion by 2015, with the next wave of growth in subscriptions expected to come from semi-urban and rural areas.[6]

Revenues in the mobile handset market are forecast to grow from US$5.7 billion in 2010 to US$7.8 billion in 2016. India is the second-largest mobile handset market in the world and is set to become an even larger market, with a forecast shipment of 208.4 million handsets in 2016.[7] It is also predicted that India will become one of the world's top five smartphone markets by 2016 and that the number of broadband connections will reach 600 million by 2020.

India's web-based economy

It is well known that the UK has the world's most internet-dependent economy. The Government of India is likewise trying to make India a web-based economy and is taking many steps to achieve this, such as the National e-Governance Plan and State Wide Area Network. Similarly, various computerization projects are being formulated for, among others, the public sector, banking, insurance and e-education, at an estimated cost of Rs 2000 billion rupees (US$37 billion). The Indian internet economy is projected to reach Rs10.8 trillion (US$200 billion) by 2016, according to a report in the Boston Consulting Group's Connected World Series study.[8] At present the Indian internet economy is growing at a rate of 23% annually, which makes it the second-fastest growing across the G-20 countries and ahead of many other developing nations in the G-20, where the average growth rate is 17.8%. It is estimated that internet video demand will account for 70% of all internet traffic in 2016.[9]

Desktop vs mobile

With the price of internet still prohibitive for many Indians and infrastructure issues that restrict access to services in rural areas, mobile has emerged as a more cost-effective and convenient internet access point. In August 2012 internet use via mobile devices was 51.63%, as against 48.37% from desktop devices. Figure 9.1 shows the comparative trends in desktop and mobile internet connection from December 2008 to August 2012.

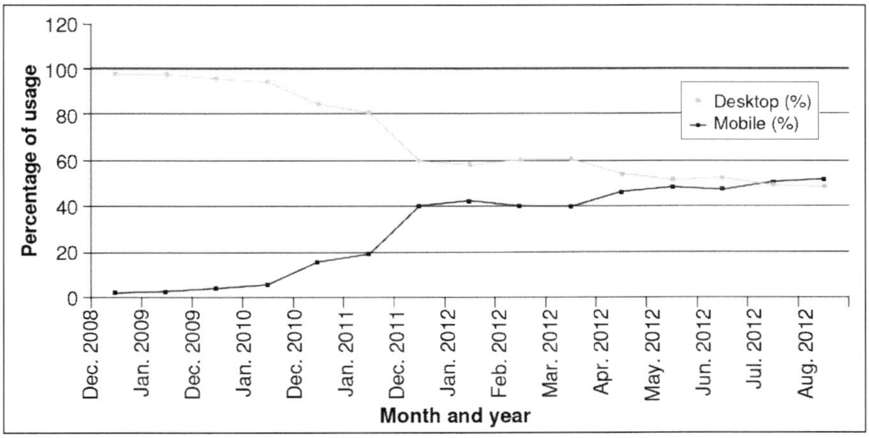

Figure 9.1 Mobile vs desktop internet connections in India
Source: StatCounter (2012)[10]

Tablets market in India

The tablets market in India is among the fastest growing in the world, witnessing an increasing growth in demand and the emergence of various players. While the market is dominated by various international brands, including Apple and Samsung, the emergence of various low-cost indigenous brands has made tablets accessible to the price-conscious Indian consumer. Of the total world sales of 45 million tablets in 2012, about 1.5 million were sold in India. According to India Quarterly Media Tablets Market Review for the third quarter of 2012, the Indian tablets market recorded sales of 1.1 million units during the quarter. In the same quarter about 135 vendors launched tablets in India. The Indian tablets market is expected to double to around 6 million units in 2013 and 23.6 million units in 2017, led by increased business adoption and the availability of Windows 8-based devices.[11]

M-education in India

It has been well established that the Indian population needs education and training on wireless devices, including mobile phones, smartphones and tablets. M-education is no longer science fiction and practical, affordable and exciting devices are available to enable learning. The development platform for m-education has become easy to implement and mobile phones and tablets are technology devices with promising applications for education. With 90% of the world population having access to a mobile network, m-education is the best way to achieve higher rates of literacy. The distinctive feature of m-learning is that it enables studying whilst on the move.[12]

India is trying to achieve a range of m-education benefits, such as on-demand access to education, instant assignments, quizzes, surveys, evaluation on mobiles, audio-video recording, feedback and various other benefits. Today, m-education has become a reality in India, and mobile learning solutions on PDA (personal digital assistant), tablet PC, hybrid phone, and other innovative mobile forms are now being delivered. Mobile learners can become producers as well as recipients of knowledge, so both 'push' and 'pull' strategies are being provided.[13] But in order to ensure a continuous and effective use of m-learning, it is essential to promote students' self-directed learning capabilities, since it is the learners themselves who are in charge of their own learning issues.[14]

The Government of India is taking many steps with public-private partnerships to develop educational e-books and e-courses, implement learning management systems and authoring tools, design games or simulation-based learning tools, provide adaptive assessment services, give test-preparation support and assist in distance tutoring and homework support.[15]

Aakash and Ubislate tablet – Government of India initiative

The search for a low-cost computing device started in India in 2002, with Simputer Trust developing and releasing a low-cost, portable alternative to the personal computer under the name Simputer – a combination of the words SIMPle compUTER which was further expanded to Simple, In-expensive Multi-lingual People's compUTER. Three years later, in 2005, the One Laptop Per Child Association Inc.

offered a US$100 laptop to Indian children in a project to provide inexpensive educational computers in developing nations around the world. But these were really expensive, in view of the Indian government's plan to distribute them to 220 million college and school students.

With this plan in mind, the Indian government promoted a tablet named Aakash,[16] and aimed to bridge the digital divide in India. Aakash, meaning 'sky' in Sanskrit, reflects the idea of reaching for the sky. The Indian government launched Aakash, the world's cheapest tablet computing device, costing Rs 2250 ($US35) in October 2011. In a pilot project it was distributed to around 100,000 students. With the government bearing 50% of the cost, the tablet was made available to higher education students for a price of Rs 1125 per unit (US$17). The Aakash tablet is also sold commercially under the label Ubislate, at a price of US$60. The latest version, Aakash2, runs on Android 2.3 and has 512Mb of memory. It will be sold commercially as Ubislate7.

In telecommunications the term 'the last mile' refers to the remotest part of a region that is to be incorporated into a network. The Aakash tablet is one such effort to connect the last person, whether in India's villages or in the metropolis, to educational technology, thereby giving Indian students a competitive edge and preparing them for competition on a global level.[17]

The Indian government's target is to make these tablets available to all 220 million college and school students. It is also proposing to make them available to students on a rental basis in college and school libraries. The tablet is preloaded with video tutorials in all disciplines, including Science, Engineering and many more. IIT Bombay has developed unique and compelling applications for engineering students on the Aakash2 platform which include 3D modelling, C++ programming, remote and collaborative training applications, robotic control and live assessment tools. IIT Bombay has partnered with the Center for Development of Advanced Computing (C-DAC) to assist with the hardware testing and logistics.[18]

Sakshat: a one-stop education portal

Sakshat is a one-stop education portal to facilitate free lifelong learning for students, teachers and those in employment or in pursuit of knowledge. The vision for Sakshat is for it to cater for the learning needs of more than 500 million people. The portal operates under a scheme known as NME-ICT which provides connectivity to all learning institutions, supplying quality knowledge modules with e-content to address the personalized needs of learners. The portal has been developed and is maintained by Indira Gandhi National Open University, New Delhi, India. It has four 'corners':

1 Student's Corner: This provides self-directed learning through virtual classes and testing services. It nurtures learners through guidance and counselling and manages scholarships under the National Merit scholarship scheme.
2 Teacher's Corner: This provides links to empowerment programmes and allows the uploading of project proposals for the NME-ICT and announcement of class schedules.
3 Knowledge Plus: This explores the world of web-based knowledge resources in the form of e-books, e-journals, study materials and lots of open education resources.
4 Interact: At this juncture, students can communicate and share knowledge with their teacher/mentor or peer group either in real time or asynchronously, through their devices.

Mobile technologies have provided unique opportunities to educators for the efficient delivery of educational materials and to support the cognitive and social processes of student learning. The Government of India has widened the learning process and extended its limits from the confines of the classroom to the internet environment. Learners can search for information on their mobile devices, synthesize the compiled information and reconstruct the information to make it into a meaningful piece of knowledge. In addition to learning content, they can receive e-mails, instant messages in text or multimedia formats, lecture notes, and audio and video files in 3G formats.[19]

The Sakshat portal provides students with educational materials that can be accessed through mobile and other portable devices. The students can communicate and interact with their peers and educators in real time and the learning process has become easy, enjoyable, time saving and portable.[20]

Conclusion

India's goal is to emerge as a knowledge super-power in the world by converting its demographic advantage into a knowledge powerhouse. The government is committed to making its population a knowledge-enabled working population through a technology-oriented economy. The mobile phone user base, now 913 million, has grown at an exponential rate and India's mobile call charges are among the cheapest in the world. The government wishes to empower the local community using technology, and to get it more involved in the education system in order to improve accountability. The government, educational institutes and various content developers have understood the huge potential of mobile devices in the field of education and many of them have successfully started to implement educational programmes tailored to citizens' needs. With the distribution of tablets, the government is providing array of m-learning apps. The Government of India Mobile Education initiative aims to accelerate the adoption of mobile education solutions; in particular, the use of mobile-enabled portable devices, such as e-readers and tablets in mainstream educational settings. In addition, various steps are being taken, such as establishing the Sakshat educational portal and the NME-ICT project, which are designed to assist both students and teachers by providing content for m-learning.

References

1 Belic, D. (2 July 2012), *World Bank: 75% of planet's population now have a mobile phone*, www.intomobile.com/2012/07/24/world-bank-75-planets-population-now-have-mobile-phone/.

2 *Global Mobile Statistics 2012 Part A: mobile subscribers; handset market share; mobile operators* (December), http://mobithinking.com/mobile-marketing-tools/latest-mobile-stats/a.

3 RNCOS – Industry Research Solutions (10 September 2012), *Mobile Learning: the future of distance education in India*, www.rncos.com/Press_Releases/Mobile-Learning-The-Future-of-Distance-Education-in-India.htm.

4 *National Telecom Policy – 2012*, www.dot.gov.in/ntp/NTP-06.06.2012-final.pdf.

5 *WDR 2012*, www.worldbank.org.

6 *Mobile Handsets: providing mobility to every Indian: industry inputs for National Telecom Policy, 2011*, emergingmarkets.ey.com/wp-content/plugins/.../download.php?id.
7 ASA and Associates (2012) *A Brief Report of Telecom Sector in India*, www.asa.in/pdf/surveys_reports/Telecom-Sector-in-India.pdf.
8 www.indiadigitalreview.com/news/indian-internet-economy-touch-rs-108l-cr-2016-boston-consulting-group.
9 *Times of India* (19 March 2012), India's Internet Growth Faster than G-20 Nations: report. http://articles.timesofindia.indiatimes.com/2012-03-19/internet/31210050_1_indian-internet-internet-growth-internet-access.
10 StatCounter (2012) *GlobalStats*, http://gs.statcounter.com.
11 *Indian Tablet Market Expected to Double in 2013; Samsung leads the pack* (2013 January), http://gadgets.ndtv.com/tablets/news/indian-tablet-market-expected-to-double-in-2013-samsung-leads-the-pack-313391.
12 Bon, A., de Schryver, T., Twinomurinzi, H. and Jordaan, D. (2012) Less (in Context) Is More (Creativity): m-learning as a short lived traveling idea at the University of Pretoria, *International Journal of Business Anthropology*, 3 (2), 75–87.
13 Woodill, G. (2012) Moving from E-learning to M-learning, *Canadian Learning Journal*, 16 (2), Fall, 34, http://c.ymcdn.com/sites/www.cstd.ca/resource/resmgr/clj/clj_fall_2012_final.pdf.
14 Liu, Y, Han, S. and Li, H. (2010) Understanding the Factors Driving M-learning Adoption: a literature review, *Campus-Wide Information Systems*, 27 (4), 210–26.
15 Pandit, V. (2012) *Transforming Learning through mEducation*, McKinsey & Company, Mumbai, g3ict.org/download/p/fileId_910/productId_224.
16 Akash Tablet (2012), www.akashtablet.com/ubislate/index.html.
17 Etoosindia (16 November 2011), *Aakash Tablet 'The First Chance to Connect the Last One'*, http://etoosindia.wordpress.com/tag/education/.
18 *DataWind's Aakash 2 Launched by President of India on World Education Day* (2012), http://ubislate.com/press-releases/AAKASH2%20LAUNCH.pdf.
19 Hussin, S., Manap, M. R., Amir, Z. and Krish, P. (2012) Mobile Learning Readiness among Malaysian Students at Higher Learning Institutes, *Asian Social Science*, **8** (2), 276–83.
20 Singh, M. (2010) M-learning: a new approach to learn better, *International Journal of Education and Allied Sciences*, **2** (2), 65–72.

this is via an app. A report by Flurry Analytics (2011) suggested that online consumers spend more time using apps (81 minutes a day) than on the web (74 minutes a day), and there are over 600,000 apps available. Health-related apps are a major growth area, with over 13,000 such apps for consumers (MobiHealth, 2011). Today, one in five smartphone users have at least one health-related app on their mobile. Exercise, diet and weight-management apps are the most common.

Mobile ways of working – increasing convenience and reducing cost in the NHS

Mobile devices are increasingly being used in the everyday work of the NHS, in particular in community settings, with demonstrable benefits and efficiencies. Until recently, most scheduling of patient visits and updating of patient records by community nurses has been a paper-based exercise, requiring nurses to travel back to their office base every day to file records. Now, in areas including Calderdale and Portsmouth, community nurses and midwives update information and schedule appointments on the go using mobiles.

Real improvements in patient convenience and in savings can be achieved with simple, everyday technologies such as SMS. Appointment reminders by SMS have been shown to halve the rate of people not turning up for hospital appointments. Providing negative test results by SMS is increasingly common. Since 2004, Isle of Wight NHS Trust has been using text messages to provide test results for sexual health clinics, without the need to come back to the clinic. There is a high degree of take-up and acceptance of SMS communication in the younger population. Public transport is expensive on the Isle of Wight and, with sexually transmitted infections disproportionally affecting people from disadvantaged groups, affordable communications can improve access to affordable healthcare (Department of Health, 2012a).

Crowdsourced cities – responding to public emergencies

Social media, often via mobile, is now joining the more traditional digital communications as a mechanism for public information and networking. In the USA, cities are using the multi-channel 311

10

Widening access and stimulating innovation through mobile health applications

Bob Gann

Going mobile for health information

Health information is one of the web's most popular topics. When faced with a symptom, three-quarters of patients who are online research their condition on the web before seeing their doctor – and 70% go online afterwards to learn more. Having found health information online, around half then take action to change their lifestyle or health behaviour, make a self-diagnosis or undertake self-care (Belliveau, 2010).

Increasingly, when people want to access online health information they use their mobile. In a study by the Pew Research Center (Fox and Duggan, 2012), one in three of all mobile phone owners have used their phone to look up health information (compared to 17% two years previously). Smartphone owners are even more likely to search for health information on their phone, with over half doing this. Not surprisingly, younger, more educated people are more likely to use their mobile to access health information online. But mobile technology also has the potential to widen access. The Pew research found that in the USA African American and Latino communities were among those more likely to use their phones for health searches.

Apps explosion

When people use their smartphones for health information, increasingly customer service platform to enable citizens to report problems such as broken streetlights, potholes, noise or pollution. Members of the public can call a contact centre, visit a website or tweet their concerns. The city authorities can then respond to resolve the problem, at the same time building a rich picture of concerns, creating 'cities built on data'.

In August 2012, Twitter was used extensively by citizens during hurricane Isaac to provide early alerts and share advice to the New Orleans 311 service, NOLA Ready. Local residents used their mobile phones to tweet real-time updates on areas of flooding, road closures, etc. It has been suggested (Johnson, 2010) that 311 is re-creating the human touch and sense of small-town community in the context of a vast metropolis, strengthening an ethos of civic accountability. The NHS Commissioning Board in England has announced its intention to develop a similar service for healthcare concerns from 2013 (Illman, 2012).

Commentators are beginning to track the role of social media in 'bio-surveillance' (early identification of emerging disease outbreaks). Increasingly, it seems that Twitter knows first. Analysis of the 2012–13 flu season showed that Twitter trends relating to flu seem to spike sooner even than Google searches, and ahead of official data sources such as the World Health Organization and Centers for Disease Control. Crowdsourced social data can be a powerful complement to more traditional means of public health observation (Norris, 2013).

Mobile in the developing world

The mobile health revolution is by no means confined to higher-income countries. In a global population of 7 billion, the total number of mobile phone subscriptions globally is 5.4 billion – and counting (UN Development Programme, 2012). Indeed, some estimates suggest that there are already more mobile phones than people in the world (Cisco, 2012). According to the UN, more people now have access to a mobile phone than to a toilet or a functioning justice system. Today there are more mobile phones in Africa than in the US. Ten years ago there were just 100,000 telephone phone lines in Nigeria – today there are 100 million mobile phones.

Around the world, m-health programmes are being applied, in particular for maternal and child health and to tackle the burden of diseases linked with poverty, including tuberculosis, malaria and

HIV/AIDS. M-health applications are improving access to timely information on emergency care, improving drug safety and improving co-operation with treatment (World Health Organization, 2011).

Simple, ingenious SMS solutions are saving lives. Here are just two examples. Thirty per cent of medicines in developing countries are fake. In Nigeria the mPedigree programme enables users to scratch off a panel on medicine packaging and text the code shown, for checking against a drug database to verify whether the medicine is genuine (Ross, 2013). In Cambodia volunteers in rural areas send text messages to regional malaria workers so that they can get treatment to the patient more quickly. Texts are then mapped using Google Earth to create 'heat maps' of the location of outbreaks.

Space to innovate

So how do government and nationally funded information services such as NHS Choices respond to the enormous opportunities of m-health and m-information? In particular, what is their distinctive role alongside an innovative marketplace in mobile tools?

In May 2012, the Department of Health in England published its information strategy (Department of Health, 2012b). The new strategy recognized the central role of mobile in the digital future of the NHS:

> At the same time as exploiting simple services like SMS text used by large numbers of people, there will be considerable opportunities to exploit the cutting-edge innovation which can come through mobile technologies. Mobiles tend to be highly personal devices, always on and always with us. They work in real time and linked to a highly specific location. The mobile industry is vibrant, creative and highly competitive. These factors create a particular climate for innovative applications by imaginative developers.

The strategy argues that government and the NHS have a responsibility for providing a core set of national information, but they cannot provide everything to everyone. There are already organizations, often in the voluntary sector, which can provide better-targeted information. In future, government will provide what needs to be provided centrally, giving others space to innovate.

> Beyond this core information provided nationally by Government and creating the right environment to support the development of an information market, the state will not intervene in the information market unless it is necessary to prevent a significant negative impact on outcomes, equalities or efficiency.

Conclusion: the role of government in today's information market

The policy position now is that the most appropriate role for government is not to create apps of its own, but instead to open up public data and content freely for reuse by independent information providers and developers. In 2013 a new NHS apps and tools platform is being launched to curate and showcase the best health and care apps in the market. At the same time, the platform will provide a developer environment, linking up creative developers with sources of data.

In summary, the role for government will increasingly be:

- to produce a core set of authoritative national content when this is best done once and shared
- to provide simple, free-to-use feeds of high-quality content and data for all to reuse and add value
- to provide opportunities for ideas stimulation and match-making
- to curate and quality-assure the best examples
- to stand back and let creative developers widen access and stimulate innovation.

References

Belliveau, M. A. (2010) Health Online: the stakes are higher, *CNN Health*, 4 October.

Cisco (2012) More Mobile Phones than Humans in 2012, Says Cisco, *BBC News, Technology*, 12 February.

Department of Health (2012a) *Digital First: the delivery choice for England's population*, Department of Health.

Department of Health (2012b) *The Power of Information: putting us all in control of the health and care information we need*, Department of Health.

Flurry Analytics (2011) *US Mobile Apps versus Web Consumption*.

Fox, S. and Duggan, M. (2012) *Mobile Health 2012*, Pew Internet.

Illman, J. (2012) NHS Info Chief Predicts US Style Revolution, *Health Service Journal*, **27**, 27 Stepember..

Johnson, S. (2010) What a Hundred Million Calls to 311 Reveal about New York, *Wired*, 1 November.

MobiHealth (2011) Consumer Health Apps for Apple iPhone, *MobiHealth News Report*, September.

Norris, B. (2013) A Look Back: comparison of flu trends uncovered by MappyHealth (analyzed Twitter tweets), Google Flu Trends, and two CDC datasets, *Social Health Insights*, 4 January.

Ross, W. (2013) Nigerian Texters Take On the Drug Counterfeiters, *BBC News, Africa*, 11 January.

UN Development Programme (2012) *Mobile Technologies and Empowerment: enhancing human development through participation and innovation*, UNDP.

World Health Organization (2011) *mHealth: new horizons for health through mobile technologies*, WHO Global Observatory for eHealth, Volume 3.

11

An education in privacy: best practices for academic libraries in the age of social media

Kate Cushon

Introduction

Throughout North America and the western world, institutions of all descriptions are finding that a digital, mobile, social media-savvy presence is not optional. Across demographics, people are embracing mobile technologies in all aspects of their lives. Academic libraries ignore this reality at their peril. They risk becoming seen as obsolete, and they risk losing user engagement.

On the other hand, embracing mobile technologies is not without risk. Financial investment in any new services or technologies is fraught, especially when many libraries are seeing their budgets slashed. The learning curve of new technologies can seem steeper with each passing year. There are also more abstract risks – these include fears about user privacy, about responsible use, about the fate of information literacy. Sometimes these risks are poorly understood, or seen as inevitable. In other cases, the risks are used as excuses to forgo mobile technologies altogether, or reasons to limit mobile technologies in ways that render them unusable.

Benefits and risks

If the adoption of mobile technologies in academic libraries is inevitable, so too are the implications of the risks involved. Academic libraries are an ideal locus for the adoption of pedagogical and social mobile technologies; so too are they ideally suited for educating university students about privacy issues and social media.

In academic libraries, institutional policies abound which address privacy concerns with regard to client information, patron records or Google analytics and other statistics-tracking software. These tend to be rather dry, hidden away in fine print on a library's website and not too likely to be accessed by the average student. As such, while they may constitute due diligence on the part of the library, they might not reach students in a relevant way.

When my institution, the University of Regina Library, began to embrace social media and mobile technology as ways to promote library use and even to provide library service, I was initially thrilled. For one thing, it brought the Library in line with current trends – in 2010 the Association of College and Research Libraries (ACRL) identified the 'explosive growth of mobile devices and applications' (ACRL, 2010, 288) as a yearly top ten trend, so we were adopting these technologies at the cusp of their eruption into mainstream academic discourse; the ACRL's list of 2012 top trends includes 'mobile environments' (ACRL, 2012, 314), suggesting that this 'trend' has staying power. On a more personal level, I was excited because I myself am an enthusiastic, if amateurish, user of social media and mobile technologies. I am aware of the ways in which social media and mobile technologies have changed our perceptions of privacy and security, but, like many users, I am rather lackadaisical about protecting my own information in these areas. With this in mind, I began to wonder about students, many of whom are younger than me, and who came of age when these technologies had already permeated their lives. I began to search for ways to ensure that our students were using social media and mobile technologies, especially those directly promoted through the Library, safely and responsibly.

Safety in the student environment

As librarians, we are also educators, and I wanted to create not merely policy but an environment where students would learn about privacy and social media as deeply as they learn how to conduct research or assess information sources. I am not a systems librarian, nor am I directly involved in my institution's adoption of social media or mobile technologies, nor am I a legal expert; this means that my focus can be on student education, rather than legal due diligence or implementation of technologies.

My initial impression was that students using social media and mobile devices are both very savvy and dangerously naïve. Most users of social media share personal information, whether deliberately or inadvertently. Users may choose to share specific pieces of personal information, as in user-created profiles that are visible to varying numbers of other users. They may also share personal information unintentionally; for example, private personal information may be hacked or inadvertently made public, or information like IP addresses may be broadcast without the user's knowledge.

Mobile technologies present their own risks. Wireless connectivity may be less secure than users are aware of, especially if they are accustomed to conducting personal business such as banking or shopping online. A history of conducting such business without assault on their personal information creates the false sense that such transactions are perfectly secure, when they may not be. Furthermore, devices that are not password protected can become lost or stolen, potentially allowing whoever finds them to access the owner's information.

As I saw it, all of these risks and more are present in individuals' personal use of social media and mobile technology. I was driven to address these risks and look into ways of educating students about them, because the University of Regina Library uses social media and mobile technologies on an increasingly large scale, both internally and in its services and promotions to students. Surely, we have a responsibility to ensure that our students are using these technologies safely and responsibly? Further, I would argue that we (as librarians) have a mandate to use and promote these ubiquitous technologies in ways that improve students' information literacy.

My preferred way to promote information literacy and sound research habits is to practise and perform the skills and behaviour I would like to promote. Rather than creating policies that students would never read and that will have little impact beyond legal protection, I aim to create an institutional environment wherein librarians and library staff engage in responsible and productive use of social media and mobile technologies. This, in turn, will demonstrate positive habits to the students who use the Library, resulting in a student body who use social media and mobile devices safely and effectively in both their academic and social lives.

Social media and privacy

Make no mistake, users of social media often do not take extra steps to protect their privacy. A 2009 study found that 'to students using Facebook, there is a false sense of privacy' (Foderman and Munroe, 2009, 37) and suggests that negative examples of harsh consequences might help. However, I would argue that negative examples abound, both in students' everyday lives (embarrassing stories or photos from friends) and in online culture (Failbook and Lamebook being examples of websites dedicated to chronicling and mocking Facebook content). A different pedagogical approach might be more effective than the suggested deterrent strategy.

A 2005 study of university students found that 'only a minimal percentage of users changes the highly permeable privacy preferences' that are the default on many social media tools (Gross and Acquisti, 2005, 71). It posits several potential underpinning factors: a signalling hypothesis ('users may be pragmatically publishing personal information because the benefits they expect from public exposure surpass its perceived costs'), an interface design explanation ('the acceptance [and possible ignorance] of the default, permeable settings') and 'peer pressure and herding behavior ... [and] myopic privacy attitudes, and the sense of protection offered by the [perceived] bounds of the campus community' (p. 80). An Australian study from 2010 gives a legal perspective, and posits six factors for young adults disclosing personal information on social media: peer pressure, signalling (creating identity), trust (of other users or of the product itself), myopic view of privacy (no sense of costs), design and relaxed attitudes about privacy (Henderson, de Zwart, Lindsay and Phillips, 2010, 6).

The situation is far from hopeless, however. The same study noting the extent of information sharing also observes: 'Although young people seem more willing to share personal information, it is important that this is not interpreted as a lack of concern about privacy, or as a complete lack of attention to the management of personal information' (Henderson, de Zwart, Lindsay and Phillips, 2010, 5), and shows how many students are at least partially aware of the risks to their privacy, even if they do not take steps to protect themselves. This means that students are highly educable in this regard, provided that the lessons do not threaten the motives they have for the sharing of information. We must teach them how to take steps to protect their privacy while still preserving the aspects of social media and mobile technology that they value.

There is a large body of literature instructing librarians on social media use, but it pays minimal attention to privacy issues. For example, an article called 'Reaching Students with Facebook: data and best practices' (Mack, Behler, Roberts and Rimland, 2007) focuses almost exclusively on the best practices from a marketing viewpoint. Another article, called 'Service Standards for the Mobile/Social Library' (Thomas, 2010), suggests a framework for planning, implementing and evaluating mobile and social technologies in the library, without mentioning privacy or security. A very thorough article on best practices for academic libraries using Facebook (Phillips, 2011) also focuses on implementation and marketing, with no mention of privacy. This is all useful, of course, especially for librarians and institutions just getting started on social media, but little or no attention has been paid to privacy issues.

In contrast to articles on implementing social media, articles on mobile devices in the library tend to have more content on privacy and security issues. In an article on the public policy end of mobile implementation, the author notes:

> Libraries should work with third-party content providers that employ privacy policies and sensible data retention practices such that information is retained only long enough to allow for the delivery of the services libraries have agreed to. Users of mobile services should be provided with clear, understandable notice as to how their personal data or location is being used by libraries or third-party content vendors. ... Libraries should realize that businesses often do not hold the same strong user-privacy values.
>
> (Vollmer, 2010, 12)

A mobile technology policy

Another article outlines the difficulties in determining ownership of mobile implementation and management on a university-wide level, which can rest with information services or other offices on campus, rather than with the library, noting that '[e]ach institution needs to consider what role the library should play in relation to mobile content and devices' (Lippincott, 2008, 4). Academic libraries are often already asked to be many things to many people, and taking on the additional burden of managing mobile technology policy may seem unwise, but the library is often the locus of mobile device use. At the University of Regina, for example, we do not consider ourselves especially early adopters, nor are we swimming in an excessive budget – yet e-book readers and tablets are loaned to students, electronic content is provided in mobile-friendly formats and librarians are provided with iPads. Mobile devices are becoming tied very closely to the services and materials provided by academic libraries and, rather than these being an unnecessary burden, I argue that libraries should fight to shape institutional policies pertaining to mobile technologies.

As a pedagogical model, I sought to create an authentic learning environment, as described in *Authentic Learning Environments in Higher Education* (Herrington and Herrington, 2006). The principles of an authentic learning environment are effective in creating an institutional dedication to information literacy, and I argue that they are also ideal for creating an institutional awareness of privacy concerns in social media and mobile technologies. Authentic learning environments have several characteristics which I decided to adapt in this endeavour:

- They 'provide an authentic context that reflects the way the knowledge will be used in real life' (Herrington and Herrington, 2006, 4).
- They include 'access to expert performances and the modeling of processes' (*ibid.*, 5).
- They encourage 'collaborative construction of knowledge' (*ibid.*, 6).

To put these principles into practice, I set as my primary objectives internal library policy and librarian education, with student education being a secondary goal. The aim is to create an environment where responsible use of social media and mobile technologies are demonstrated constantly.

On the policy end, I am creating strategy documents together with a colleague who has been instrumental in our social media and mobile activities. These documents are being developed in consultation with the University's Copyright Officer, whose office is in the Library, to ensure that the policies are sound from a legal perspective. In addition to outlining the strategic plans for developing social media outreach and mobile accessibility, the responsibilities of the institution are set out. These constitute such responsibilities as not allowing confidential or sensitive institutional information to be compromised through social media, but the document does not address more subtle aspects of responsible social media use, nor does it address individual librarians' social media or mobile technology use. Creating library policy is an important undertaking, because it helps to foster an institutional sense of the importance of social media and mobile technologies in the library – policy must be approved by the Library Leadership Team, and then presented to the Librarian and Archivist Council, so that new policy tends to permeate the consciousness of all library staff and administration.

Having said that, policy tends to be directly consulted only when absolutely needed, and is otherwise ignored by librarians in their day-to-day activities. In order to create an authentic learning environment for educating students about responsible social media and mobile technology use, I wanted to be sure that librarians and Library staff were themselves experts in these areas. To that end, I sought out and implemented a variety of tried-and-true approaches through my role on the Staff Training and Professional Development Team – I purchased webinars for all staff on mobile technologies and social media, I directed discussion after these webinars to include awareness of privacy issues, I hosted brown-bag discussion sessions on our assumptions regarding privacy laws and the protections (or otherwise) offered by third-party companies and I conversed with my colleagues about all of these issues. I also encouraged my colleagues who interact directly with students (through library instruction, the reference desk and the borrowing desk) to include education about privacy issues with students.

When going to put some of these findings into practice, I was supported by a number of individuals and groups already in existence. The Copyright Officer is an expert in legal matters and privacy law. We have an active E-book Working Group and a Mobile Apps Working

Group (now no longer in existence). We have two systems librarians, a website administrator and a library systems support staff member. There is also some institutional support for the kinds of policies and education that I was proposing. The Office of the University Secretary promotes 'Access to Information and Protection of Privacy Month' every September, with educational seminars and roundtables offered for all University faculty and staff. Additionally, University counsel holds regular information sessions for University staff on our rights and responsibilities under privacy law.

Conclusion

Academic libraries should be encouraged to take ownership of educating students on the privacy issues of social media and mobile technologies. By definition, our users are already participating in learning or teaching; if they are doing so in an environment where responsible mobile technology use and social media participation are consistently demonstrated, they will pick up on it, and our students will become excellent stewards of their own privacy.

References

ACRL Research Planning and Review Committee (2010) 2010 Top Trends in Academic Libraries, *College and Research Libraries*, 71 (6), 286–92.

ACRL Research Planning and Review Committee (2012) 2012 Top Trends in Academic Libraries, *College and Research Libraries*, 73 (6), 311–20.

Foderman, D. and Monroe, M. (2009) The Impact of Facebook on Our Students, *Teacher Librarian*, 36 (5), 36–40.

Gross, R. and Acquisti, A. (2005) Information Revelation and Privacy in Online Social Networks. In Atluri, A., De Capitani di Vimencati, S. and Dingledine, R. (eds), *Proceedings of the 2005 ACM Workshop on Privacy in the Electronic Society*, ACM.

Henderson, M., de Zwart, M., Lindsay, D. and Phillips, M. (2010) Legal Risks for Students Using Social Networking Sites, *Australian Educational Computing*, 25 (1), 3–7.

Herrington, A. and Herrington, J. (eds) (2006) *Authentic Learning Environments in Higher Education*, Information Science Publishing, 321.

Lippincott, J. (2008) Mobile Technologies, Mobile Users: implications for academic libraries, *ARL Bimonthly Report*, 261, 1–4.

Mack, D., Behler, A., Roberts, B. and Rimland, E. (2007) Reaching Students with Facebook: data and best practices, *Electronic Journal of Academic and Special Librarianship*, 8 (2), n.p.

Phillips, N. (2011) Academic Use of Facebook: building relationships with students, *The Journal of Academic Librarianship*, 37 (6), 512–22.

Thomas, L. C. (2010) Service Standards for the Mobile/Social Library, *Journal of Web Librarianship*, 4, 449–51.

Vollmer, T. (2010) There's an App for That! Libraries and Mobile Technology: an introduction to public policy considerations, *ALA Office for Information Policy, Policy Brief*, 3, 1–14.

12

QR codes as teaching tools

Keiso Katsura

Introduction

The history of QR code development in Japan

The QR (Quick Response) code, a matrix-type two-dimensional bar code, was invented by researchers at Nippondenso Co. Ltd in co-operation with Toyota Central R&D Labs Inc. in 1994. Both organizations are located in Aichi prefecture, central Japan (Figure 12.1). 'Nippon' means Japan and 'denso' is the combination of 'den' (electronic) and 'so' (equipment).

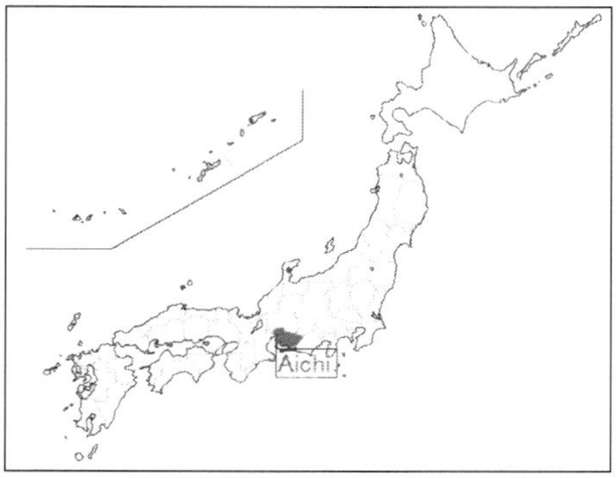

Figure 12.1 The birthplace of the QR code in Japan

Nippondenso's original English name was Nippon Electronical Equipments Co., Ltd. Nippondenso changed its name to Denso in 1996. Denso established its subsidiary company, Denso Wave Incorporated by separating its QR code development division in 2001. Since then, Denso Wave has been responsible for the QR code.[1]

As to who invented the QR code, this has been a subject of discussion outside Japan. Things will become clear if it is pointed out that the Japan Patent Office (JPO) accepted a patent application from five applicants in 1994. Three were from Nippondenso and two were from Toyota Central R&D Labs Inc. These applicants were authorized as the patentees in 1999.[2] The JPO patent number is 2938338. In recent years the QR code has become popular outside Japan.[3] Table 12.1 shows the chronology of the development of the QR code in Japan.

Table 12.1 Timeline of QR code development in Japan[4]

Year	English name	Development
1949	Nippon Electronical Equipments	A car parts manufacturing company established by separating it from Toyota in Aichi, Japan
1960	Nippondenso	Name change to Nippondenso
1992		Started QR code research
1994		Developed QR code in collaboration with Toyota and applied to the Japan Patent Office for registration
1996	Denso	Name change to Denso
2000		QR code was authorized as ISO/IEC18004
2001	Denso Wave	Denso Wave established as a subsidiary company responsible for QR code, etc.
2002 to present		Sharp, a Japanese electronics company, installed QR code recognition software in mobile phones

QR codes and mobile devices

Two advantages of the QR code are that it avoids the need to key in URL addresses, especially longer and complicated ones, and that it can be read by mobile devices. Since the QR code patent was treated as licence-free or public domain, it was easy for mobile phone and printer manufacturers to install the related software in their mobile phones and printers. In 2002 Sharp, a Japanese electronics company, released mobile phones on which the camera was equipped with a reader to scan QR code images.[5] Other Japanese mobile phone manufacturers then followed suit. By 2007, when the seminar 'Cell Phone Teaching in Japan' was held at the Media Zoo at the University of Leicester, Nokia had already installed QR code scanners in some of its mobile phones.[6]

The QR code was originally intended to improve the Kanban supply chain system between Toyota, a company that manufactures automobiles, and Nippondenso, which manufactures car parts. However, once mobile phone camera applications allowed everyone to read QR code, it became a popular utility.

QR codes and label printers

The rapid diffusion of QR codes was also triggered by the growth both of free websites that generate QR codes and of QR-code-compatible printers, including label printers. Some Japanese label-printer manufacturers, such as Brother Industries, Ltd., have developed QR-code-compatible printers.[7] Figure 12.2 shows two pictures (a and b) of the Brother P-touch 2430PC printer.

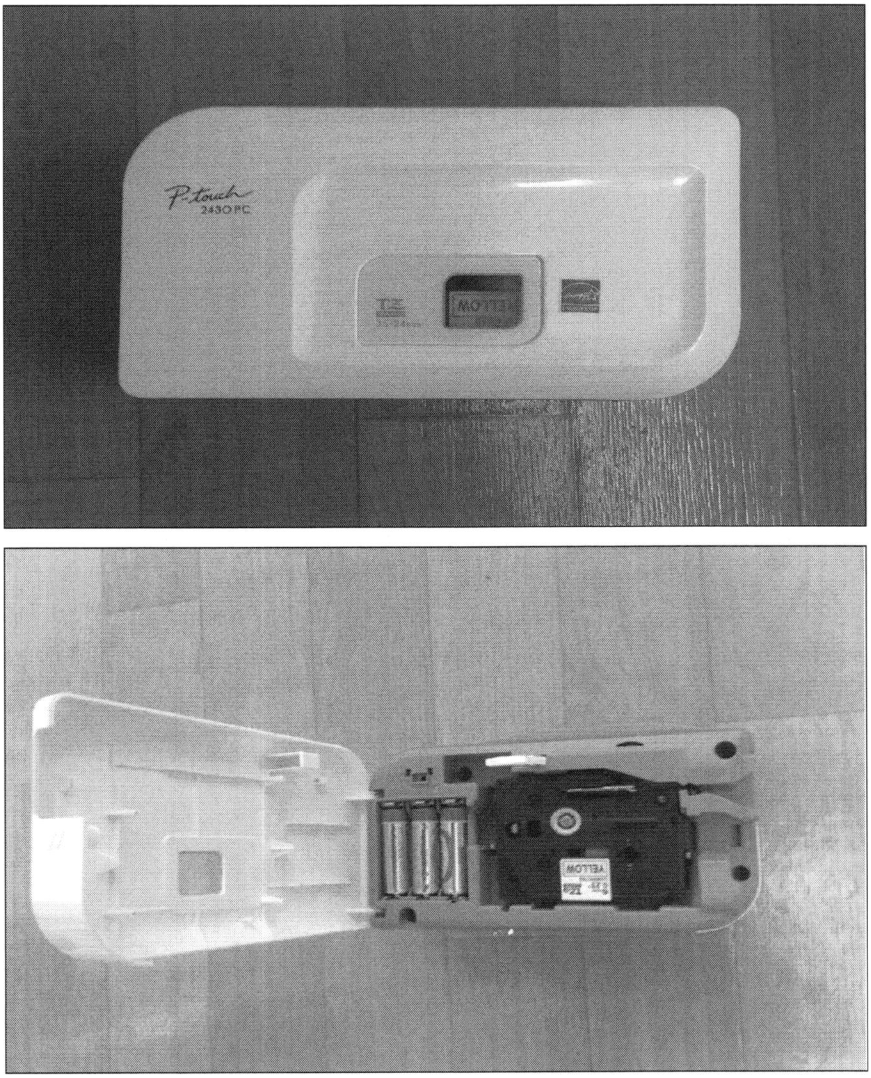

Figure 12.2 Brother P-touch 2430PC (a) exterior, (b) interior

The strong point of the QR code is that it can be applied to all media. It can be printed not only on paper but also on any other medium, such as clothes, walls and even stones. A group of people wearing black and white clothes can constitute a QR code, too. NFC (near-field communication), on the other hand, which can also read URL addresses, requires a metal chip. If a label printer is available, it is easy to replicate the content of a QR code. Label printers include software that generates QR codes and they can print codes that have been copied and pasted from other sources.

QR codes consist mainly of black and white. But from a visual viewpoint, labels with a yellow background are more effective than those with a white background. The colour and size of the label tape supplied by label-printer manufacturers can be specified.

Case study of QR codes in teaching

In educational settings, QR codes are more useful when applied to teaching tools than to teaching content. This case study reports on the use of QR code in a Library and Information Science (LIS) course at Miyagi Gakuin Women's University, Sendai, Japan. In Japan, most students bring their mobile phones equipped with QR scanner into the classroom. The LIS students also bring their mobile phones and/or smartphones to class.[8] Some students may bring tablets or mini-tablets that include QR code applications and can access the internet. However, costs and battery life mean that it is not suitable to use mobile devices in the classroom all of the time. In the near future, all students on the LIS course will use mini-tablets with QR code applications that will be loaned to them.

QR codes as tools for communicating with students

In the very first class, the enlarged QR code with the URL for the course is shown on the monitor at the front of classroom. The more than 150 students in the class immediately scan and acquire the code using their mobile phone cameras. In subsequent classes students can report class attendance without inputting the URL each time. The students can also receive any information about the course anytime and anywhere. There is also a laminated label on the office door of the LIS professor with both the QR code that was shown in the first class and an NFC chip containing the same URL address (Figures 12.3 and 12.4).

Figure 12.3 QR code and NFC chip on professor's office door

Figure 12.4 Components of the QR code and NFC chip label

A simple web page to which the QR code is linked includes the following content:[9]

Figure 12.5 Real-time display schedule, run by a CI program

1 links to the professor's real-time schedule (Figure 12.5)
2 links to student class attendance report system
3 course bulletin board
4 recent library job announcements
5 links to mobile OPACs
6 links to LIS-related websites
7 links to RSS feeders with information about recent library articles and events
8 links to real-time questionnaire system used in class
9 other links and content that were recently added.

QR codes on the library shelves as a subject guide

This paper also suggests further sophisticated applications of QR codes to library services, such as the use of QR code on library shelves, known as a Shelf-based Subject Guide (SBSG). If the URL of OPAC search results is included in the QR code, when they are standing among the library shelves students can acquire bibliographic information about books on particular shelves arranged by call number, regardless of whether or not the books are present on the shelves. The following are advantages of affixing this QR code to the shelves:

1 It is not necessary for students to plan complicated searches, as the QR code for them is set up by the professor in advance.
2 The professor can add comments and reviews to the list of books on the shelves. Teaching staff can play the role of 'subject guardians'.
3 Not only librarians but also teaching staff can prepare QR code labels and attach them to the shelves.

4 The staff can add to the QR code not only book lists but also lists of related articles, web resources and subject headings extracted from the Library of Congress.
5 The staff can change links from local library records to commercial ones, such as Amazon.
6 Students can identify the books present and not present on the shelves by comparing the list and the books.
7 Students can check the location marks of books that are shelved in other locations although they have been assigned the same classification numbers. For example, books with the same classification number may be stored in different places in the library, such as central stacks, closed stacks, reference collections, large-size books or rare book collections. Students are sometimes confused by the library's book location system.
8 RSS feeds can always automatically update the bibliographic information in the lists.

The QR codes shown in Figures 12.6 and 12.7 on the following page list the books on the corresponding shelf. These books are on the subject of information retrieval. NDC (Nippon Decimal Classification) is a nation-wide Japanese library classification system. NDC007.58 is equivalent to DDC025.524. The lists connected to the QR codes can include books on a specific subject in the local library and web resources on the same subject.

Figures 12.8, 12.9 and 12.10 (page 109) reflect the content of the QR codes in Figures 12.6 and 12.7.[10] In Figure 12.8, the books in the list correspond to those on the shelves. The content searched for by the professor was reorganized using an HTML editor. Figures 12.9 and 12.10 show additional reference resources on information retrieval. In both examples the RSS-based content was fetched by JavaScript.

Additional technological matters

Further development of QR code

QR code is still being developed by Denso Wave. The company has recently deployed iQR code, which can be rectangular in shape, and SQRC (Security QR code).[11] Further technological innovations of QR code will be considered.

Figure 12.6 Yellow QR code tape on a bookshelf

Figure 12.7 QR code tapes generated by a label printer

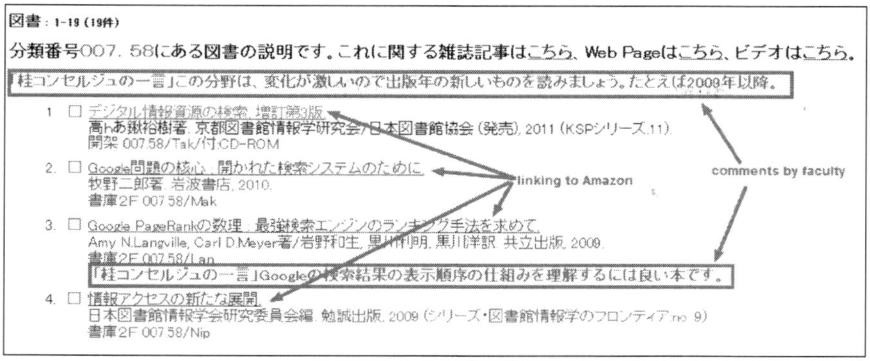

Figure 12.8 Local OPAC search results linked to the QR code

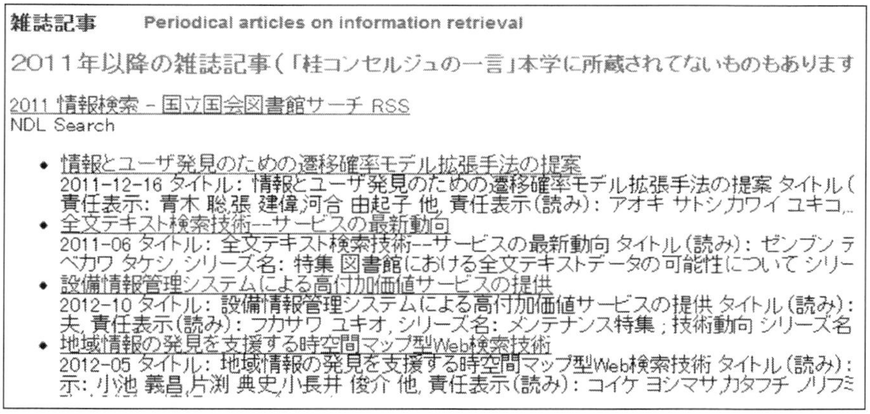

Figure 12.9 Articles extracted from the Periodical Index of the National Diet Library

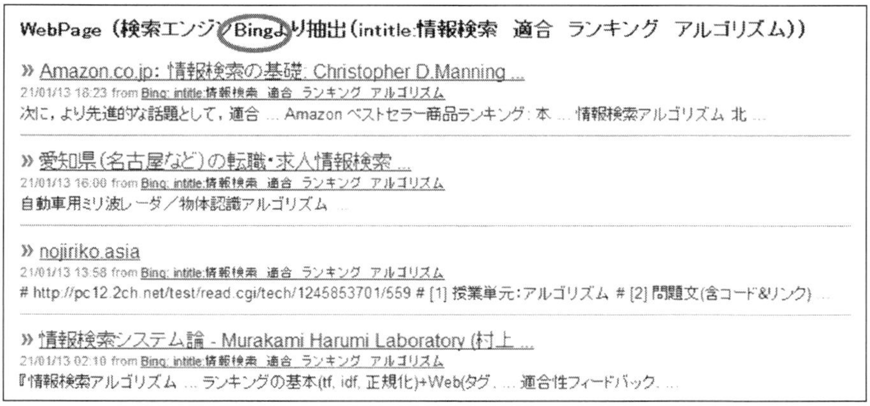

Figure 12.10 Web pages retrieved from Bing and processed by RSS

Duplex use with NFC

Recently some questions have been examined regarding whether NFC can replace QR code or whether the two can be used in parallel.[12] If the number of mobile devices that have NFC capabilities increases, this may become a reality. Kimtag, a London-based company, has released the TwinConnect tag, which includes both QR code and NFC.[13]

Notes and cited references (with QR codes)

1 The address of Denso Wave's English web page: www.denso-wave.com/en/.

2 They (all males) are Hara, Masahiro (Nippondenso), Watabe, Motoaki (Nippondenso), Nojiri, Tadao (Nippondenso), Nagaya, Tadayuki (Toyota Central R&D Labs) and Uchiyama, Yuji (Toyota Central R&D Labs). The English spellings of their names were advised by the Denso Wave Incorporated by email.

3 A survey of QR code use in the UK shows that the use of QR code scanning grew by 700% from January to July 2011: www.qrmediacodes.com/qr-news/214-qr-code-uk-survey.html.

4 *Denso 50 nenshi: shiryohen*, Kariya: Denso, 2000, 19–20. (The main title of this book can be translated as 'A Fifty Years History of Denso.')

5 Muramatsu, T., Suwa, A. and Miki, K. (2003) The QR Code Technology of a Cellular Phone with a Camera, *Sharp Technical Journal*, **87**, 49–53. (Text is Japanese with abstract in English and Japanese.)
The PDF full-text version is available at:
www.sharp.co.jp/corporate/rd/22/22-5-3a.html.

6 In the seminar, one participant from Italy showed that her Nokia cell phone camera had a QR code reader:

www2.le.ac.uk/departments/beyond-distance-research-alliance/dissemination-activities/seminars-symposia-and-workshops/2007/mobile_technologies?searchterm=None.

7 The Brother Label Printer, PT-2430PC Labeler, can run on both 110–240 volts and also 6 × AA batteries.
The address of the product demo page:
www.brother-usa.com/pt-2430pc-labeler/#.UO_LFOSExcw.
The headquarters of Brother Industries is also located in Aichi, Japan.

8 In a survey conducted on 20 September 2012 in the LIS class, 100% of 97 students were carrying their mobile phones, and 41 (42%) students had their smartphones. Ten students were iPhone owners and 30 students were Android owners.

9 The web address of Keiso Katsura's Mobile Portal:
keisok.sakura.ne.jp/i/a2.htm.

10 The web address of NDC007.58:
keisok.sakura.ne.jp/mgundc/ndc007.58.htm.

11 An introductory page on iQR in English is:
iQR is the new QR code:
2d-code.co.uk/iqr-qr-code/.
The URL of an explanatory page about SQRC by Denso Wave (text in Japanese) is:
www.denso-wave.com/ja/adcd/product/software/SQRC/sqrc.html.

12 A typical discussion on QR codes vs NFC can be found at the following site: www.newyyz.com/ntgsite/tag/nfc/.

13 The Kimtag website is: Kimtag: a connection hub: kimtag.com.

13

Making sure to remember what we already know: ensuring e-reading innovation works

Anne Hewling

Background

> We often say in mobiles-for-development that today most people in the developing world will make their first phone call on a mobile, and have their first experience of the Internet on one, too. Perhaps children, in the not-too-distant future, will have their first experience of reading on an e-reader?
>
> Ken Banks, 30 January 2012[1]

The road to hell is paved with good intentions. In the 1960s and 1970s the fields of Africa were littered with tractors and other mechanical toys sent by well-meaning donors who focused on what they thought they knew – that mechanical farming increased yields – whilst forgetting that such machines need parts and fuel to do so. They also cheerfully ignored the labour force changes required for a socio-economic environment that would encourage the economic growth they envisaged (and thus reduce the poverty which was the target of their endeavours). Some would argue that little has changed. One Laptop Per Child (OPLC) and other technology initiatives have been accused by some of replicating the same paternalistic mistakes of the 1960s and 1970s. Others, like Ken Banks, above (M-Libraries keynote speaker in 2009), see the affordances of new technologies as powering almost irresistible change.

In practice it is extremely easy to create electronic text, find a network to transmit it, then locate a device which can store and reproduce it. Every week, or so it seems, new devices, new software and better access are touted by a worldwide market that has already delivered 6.5 billion mobile subscribers. Saturation phone ownership numbers have been reached in countries labelled as 'developed' and phone ownership rates in the world generally indicate that gadgets like smartphones, or devices with similar capability, are now the technology of choice for much of the world's population. At the time of the M-Libraries conference (September 2012), keynote speaker Steve Vosloo reported that 77% of mobile connections were in developing countries. Other sources report that one billion women in low-income countries own a mobile phone of some kind. And, although the mobile playing field is not level, solutions are now available to previously insoluble related problems such as the non-availability of electricity, absence of service providers, remote physical location, etc. Solar panels and wireless radio transmitters and satellites mean that access is feasible from locations that just a few years ago seemed destined to remain perpetually out of range. Marketing and a global drive towards raising levels of education, perceived as a certain route to economic development and success, are encouraging the development of more and more online education solutions. The potential for growth seems almost limitless, particularly in places where traditional solutions are not possible, not working or in need of renewal. The potential for development of e-reader solutions in such a context is huge. So too, of course, in the rush to be leader in the new world order, is the potential to get it wrong.

Traditionally, access to reading material is brokered to a greater or lesser extent by gatekeepers in the form of editors and publishers and, for education, by teachers, school boards and other guardians of the curriculum like local and national education authorities. There have always been disagreements over what is suitable, or not, but, over time, consensus and rebellion have balanced a status quo. This is now being challenged by the development of technology that allows individuals and interest groups to perform tasks previously owned and managed by status quo gatekeepers.

This chapter begins by situating the e-reader – both tool and user – in the context of reading and literacy within the changing technological context. It then considers generally the issues to be addressed in ensuring that the e-reader device meets the needs of the e-reader user in this new context. Finally, the chapter proposes a simple framework to guide those developing resources for both the tool and the user in an educational context.

E-reading in the changing technological context

Literacy is highly valued in the modern world. Investigation shows that those with at least basic literacy gain benefits in health, wealth and life chances generally. Reading is a key part of that literacy. The ability to read well is dependent on practising reading. In turn, this requires access to reading material, both formal and informal. Beyond mastering the physical processes of reading is the psychological element of becoming 'a reader', i.e. someone who sees themselves as a competent practitioner. The physical and psychological elements of being a reader require nurturing, and this has traditionally been done by teachers, librarians, writers and publishers. These roles are well established and understood and their rights and responsibilities are tacitly, if not explicitly, agreed.

The reader's degree of success at being a reader goes a long way towards success in education, and thus in employment and life more widely. Any change to the status quo therefore has potentially serious repercussions for the reader and their wider society. The arrival of computers impacted upon reading in a number of ways. Ideas of what was assumed to be a new electronic literacy were documented by authors such as Nancy Kaplan (1995), who theorized dealing with hypertextual, non-linear reading. However, with time and daily practice, many of the issues that Kaplan identified have been confined to specialist areas of online reading. The potential (threat) of the hypertext 'book' has given way to a more literally translated version of the paper-bound book in the form of the e-reading device. With this, the tradition of linear text, in sentences, paragraphs and chapters, is retained but material is delivered on a book-shaped device by a typewritten, electronically generated typeface. In fact, manufacturers have gone to great lengths to replicate the physical aspects of the traditional reading process whilst adding the potential to carry a whole lifetime's reading on a single book-sized device (Breckenridge, 2012). Furthermore, the texts themselves are given new power in that they can be formatted in multiple ways. They are not even confined to being housed in the physical body of the e-reader but can be reproduced for portability on mobile devices via e-reader software that is downloadable onto any internet-enabled phone or other mobile device. The library and bookshop are where the reader is 24/7. The reader no longer needs the intermediary assistance of publisher, librarian or bookseller, unless they so choose. Nor does the author. The potential exists for all to access previously inaccessible information and resources, to visit and create new worlds.

In many respects, therefore, it is hard to see how the e-reader (device) can be viewed as other than an enhancement to the e-reader (person) experience. In fact, the implications of this change are wide. Both social and physical processes and practices are impacted upon and changed boundaries need to be recognized and addressed.

At the highest level some issues are worth thinking about, not because much can be done about the inevitability of change inherent in the human condition, but because there is a need to acknowledge the unpredictability of the change that actually happens (Conley and Wise, 2011). Fewer than 20 years ago it seemed quite likely that the printed book would vanish or, at best, be reduced to a print-on-demand commodity. Many pages of *The Bookseller* magazine (www.thebookseller.com/) were filled with articles on the viability of print-on-demand machines, and this author, then a bookseller, was warned that the small independent bookshop had no future. The ubiquitous 'everyone' would, it seemed, be reading hypertext, non-linear material on computers – then envisaged as smaller versions of desktop machines; e-ink technology remained a dream. Few were predicting the prevalence of e-readers or just how many books would be read on mobile phones or tablets. The key factor here is that huge change happened, and not exactly as predicted. The learning point for the future is that any medium- to long-term planning for the use or development of current e-readers needs to allow that five or ten years will bring change, but it may not be as currently envisaged. But whilst it may be impossible to predict the future, it is possible to ask questions and take action now to ensure that future practice is built on solid foundations. Sound pedagogy and principles can go a long way to future-proofing the educational use of e-readers. Establishing specifications for infrastructure, good practice guidelines for users, authors, etc. is part of this process. Aspirations and expectations will move over time, and expecting change is essential but less daunting when approached from a solid baseline. Important questions to guide such a baseline are contained in the framework at the end of this chapter.

An e-reader device to meet user needs

On a more mundane level, ensuring success in developing e-reader activity requires attention to two principle areas: the infrastructure and the experience. Fundamental to infrastructure are issues of sustainability and whether the use (possibly the first introduction) of e-readers makes sense in terms of resources available now and for the foreseeable future. Leaving aside whether or not anyone would want the option to do so, it is technically possible to run a printer or photocopier from solar power in the middle of the Kalahari desert. However, dust and the lack of locally available parts and servicing mean that it is unlikely to make sense to do so. Simply put, just because you can do something, as the saying goes, does not necessarily mean that you should do it; a feasibility assessment for any new e-reader initiative is probably essential.

Issues related to the experience of e-reading development can be sub-divided into three inter-related but distinct areas: the tools, the people and the material.

Ken Banks (2012) asks an important question which helps to clarify the issues:

> So, what makes the Kindle so special?
>
> - It's light, relatively rugged, and mobile
> - Ten days reading time on one charge
> - One month 'standby' time between charges
> - Solar panel cover option removes the need for mains charging
> - Built-in dictionary and thesaurus
> - Display can be read in bright sunlight
> - Internal storage for up to 200 books
> - No need for the Internet once books are loaded
> - Text-to-speech for illiterate/semi-literate users
> - Costs continue to come down
> - Remote delivery of books and materials (local wi-fi permitting).
>
> (Banks, 2012)

Consideration of tools requires an assessment of potential devices in relation to previously stated infrastructure concerns and local physical conditions. The first Worldreader (www.worldreader.org) project with Kindles, in Ghana, failed to anticipate both just how many broken screens it would endure and that repair would involve the Kindles' being sent back to the US. Subsequent stages of the project and a more recent project in Kenya (Poverty Matters Blog, 2012) have made sure to use a more robust device and to secure more local support.

Consideration of the people element of e-reading projects requires breaking down any danger of 'them and us'. First and foremost, this implies a need for mapping the differences in expectations and aspirations for the use of e-readers. Those implementing introduction or development of e-reader use are likely to be at least one technology generation away from those who are expected to pick up and use those e-readers. It will be essential to know whether these two generations share views on what ultimately will constitute success for the project. It is also highly likely that scepticism about the benefits of technology will be greater amongst those on the receiving end of the project, and the potential of the initiative is likely, initially at least, to seem less obvious to them too. Feelings of technophobia or cultural distrust may arise and, when a project is donor funded, technical skills will likely be significantly higher amongst the implementers than amongst those who will actually carry forward the criteria for a successful outcome of the project. This cannot be allowed to create any further barrier to implementation.

Consideration of the material to be delivered by any e-reader development requires recognition that it is nothing without the other elements – people and tools – and a certain degree of compromise may often be required, although this in turn raises further issues. The Worldreader projects mentioned above use donated texts from mainstream western publishers and thus, immediately, raise issues of ownership of materials, ideas and attitudes – much as have some print textbook initiatives in the past. A considerable advantage, though, is that local material can be accorded suitable priority, as has happened in Ghana, although according to language and notation there must be some adaptation of software to allow non-standard characters.

Material can, of course, also exist independently of the other two elements, i.e. as books or other general publications. In fact e-readers here outpace more traditional materials such as printed-word books, by way of their extraordinary properties outlined by Banks above – in particular, the potential for instantly broadcastable two-way interaction that is not single- (English) language specific and does not require permission from or validation by traditional gatekeepers. I, for example, can easily publish my own material in my mother tongue regardless of what that language is or the nature of the material. I can develop it alone or alongside others of my choosing and accompany it with images, sounds or any other multimedia elements that my chosen format will support.

Framework for developing resources in an educational context

The development of new e-reading projects holds enormous promise for offering reading access, interaction and practice to many who cannot presently avail themselves of these resources (Larson, 2009). The constraints on development are few and the potential benefits in terms of options for localization and globalization are considerable, provided that care is taken in the planning and execution of such initiatives. Below is an outline framework to assist this process. It is not a definitive structure, but must be seen as a work in progress, given the nature of change as outlined above.

Framework for successful development of e-reader projects

<table>
<tr><td>Big questions</td><td>
<ul>
<li>Who owns this project?</li>
<li>Is an e-reader project the best way of meeting user needs? If not, why is the project needed? How will it be useful? What is the pedagogical support for this project?</li>
<li>Is this project sustainable?</li>
<li>Is it dependent on any of the following:
<ul>
<li>Named individuals? (Could they be replaced if necessary?)</li>
<li>Donated resources of any kind? (Could they be replaced if necessary?)</li>
<li>Non-recurrent budgets? (Could they be replaced if necessary?)</li>
<li>Does this project require or assume any reduction in resources elsewhere, e.g. library budgets/cuts? What plan has been made for resistance/failure of any/all of these assumptions?</li>
</ul></li>
<li>Do you have buy-in from:
<ul>
<li>Potential users?</li>
<li>Potential authors/materials writers?</li>
<li>Curriculum gatekeepers/school boards/local education authorities?</li>
<li>Parents and local community?</li>
</ul></li>
<li>Replicability and/or cost sharing potential?</li>
</ul>
</td></tr>
</table>

	Infrastructure and devices	People	Material/content

Little questions	• Device
	  • Physical issues – strength? Resistance to likely damage? Repair and replacement arrangements?
	  • Usability issues – shape, size, design, buttons, menus, mouse?
	• Connections
	  • Institution only/open to wireless?
	  • Wireless access arrangements?
	  • Hardware?
	  • Costs?
	• Management of:
	  • devices
	  • users
	  • service/other contracts?
	• Updating/virus and malware issues?
	• Budgets and finance, fixed and recurrent?
		• Who are target population?
		  • Familiarity with devices?
		  • Issues of ownership/use of devices/storage/ liability for damage and misuse?
		  • Contractual arrangements?
		  • Access to internet?
		  • Choice of material on device?
		  • Code of conduct?
		• Current literacy levels and practices?
		• Cross-consistency with new materials?
		• Only readers?
		• Readers and authors?
		• Who can author?
		• Authors' contracting arrangements and intellectual property rights management?
		• Cultural and social issues?
		• Training and helpdesk roles and activities?
			• Who will/can author materials?
			• What will be the nature of content? Prescribed formats?
			• Curriculum/content integration – who decides?
			• Who can commission/authorize/ distribute content?
			• Are controls required on all content to be added?
			• User rights to add/delete content?
			• Quality control and licensing of new material? Of user-generated content?
			• Cultural literacy issues?
			• Ownership of content – short and longer term?

Note

1 Accidental appropriate technologies – a blog post by kiwanja, www.kiwanja.net/blog.

References

Banks, K. (2012) Accidental Appropriate Technologies, *Build It Kenny, and They Will Come …* (January), www.kiwanja.net/blog/2012/01/.

Breckenridge, K. (2012) E-readers the Answer to Textbooks, *Mail & Guardian*, 19 October, http://mg.co.za/article/2012-10-19-e-readers-the-answer-to-textbooks.

Conley, M. W. and Wise, A. (2011) Comprehension for What? Preparing students for their meaningful future, *Theory into Practice*, **50** (2), 98, http://dx.doi.org/10.1080/00405841.2011.558411.

Kaplan, N. (1995) Politexts, Hypertexts, and Other Cultural Formations in the Late Age of Print, *Computer-mediated Communication Magazine*, 2, www.ibiblio.org/cmc/mag/1995/mar/kaplan.html.

Larson, L. C. (2009) Digital Literacies: e-reading and e-responding: new tools for the next generation of readers, *Journal of Adolescent & Adult Literacy*, 53 (3), 255–8.

Poverty Matters Blog (2012) 'E-readers Kindle Enthusiasm for Learning among Children in Kenya', *Guardian*, www.guardian.co.uk/global-development/poverty-matters/2012/oct/26/e-readers-kindle-learning-children-kenya.

14

Text reference service: ideas for best practice

Lili Luo

Introduction

Today's technology offers new ways for librarians and users to interact and also provides new opportunities for libraries to deliver services collaboratively to wide and diverse audiences. Text messaging, or texting, is one such technology that libraries are embracing. Texting allows the exchange of text messages on mobile phones. Each message can be up to 160 characters (words, numbers or alphanumeric combination) in length when the Latin alphabet is used. Texting has become an increasingly significant means of communication and social activity in people's lives. The Pew Research Center's Internet and American Life Project reported that 83% of American adults own mobile phones and 73% send and receive text messages. Young adults are the most frequent texters – mobile phone owners between the ages of 18 and 24 exchange a daily average of 109.5 messages (Smith, 2011). Recognizing the popularity of texting as a means of communication, a growing number of libraries are offering texting-based reference service (text reference service) for their users to seek information and assistance from librarians.

In 2010 the School of Library and Information Science at San Jose State University received a grant from the Institute of Museum and Library Services to study the practice of text reference service, investigating how texting is being used as a new means of providing reference service and how text reference service can fulfil users' information needs and engage new users, such as teenagers, the fastest-growing group of individuals using text messaging. The project draws upon the rich pool of data available via My Info Quest (MIQ), the first collaborative text reference service participated in by over 20 multi-type libraries across the United States. MIQ was launched in July 2009. It is self-organized and managed by volunteer member librarians. During the two-year project period the team conducted a number of research studies to examine text reference service from the following perspectives:

- types of information needs of text reference users
- user perceptions and use of text reference service
- competencies requisite for text reference service.

This chapter presents the findings of the project and discusses their implications. The in-depth analysis of text reference practice will help in generating best practice guidelines and therefore lead to an enriched view of texting's affordance as a means of reference service. Libraries will be able to make more informed decisions about service planning and implementation and, ultimately, to deliver a service that both aligns with the library's mission and optimizes the users' experience.

Text reference service: how it works

Text reference service is defined as reference service provided via the exchange of text messages on mobile phones. Library users text their questions to a phone number provided by the library and answers are sent to them in the form of text messages. Librarians usually use a computer application to monitor user questions and compose the responses. The connection between computer and text messages is supported by gateway software such as Text A Librarian and Altarama SMS Reference. Figure 14.1 illustrates the text reference process.

Lippincott (2010) lists some broad questions to consider when planning text reference service:

- What is the current state of mobile device deployment at your institution or for your population?
- What are your goals for providing service and what are your strategies?

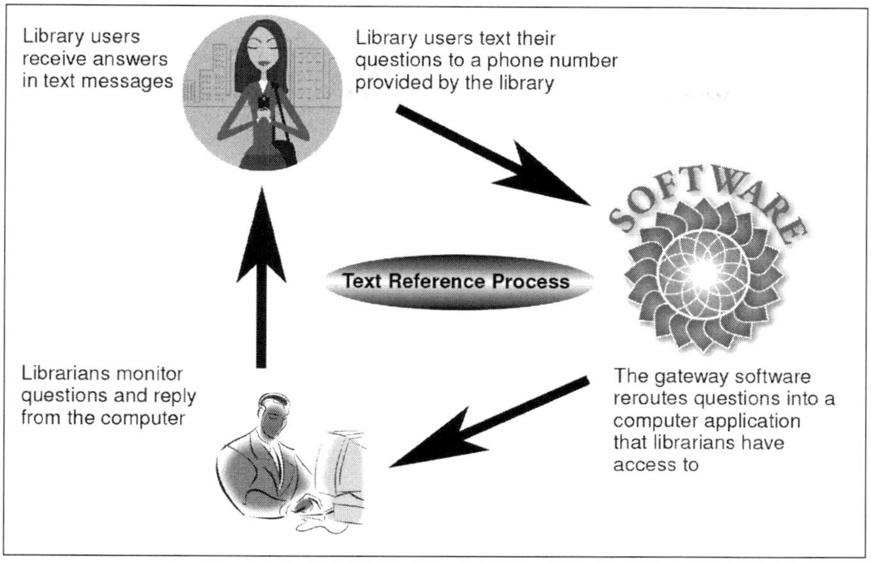

Figure 14.1 The text reference process

- Whom should you work with in your institution or service area?
- How will you know if you are successful?
- What is your strategy for the next two to three years?

These questions, according to Lippincott, are 'fundamental questions that can assist libraries in the framing and planning process' for text reference service (2010, 2). They can help libraries to assess various service needs and therefore set objectives and policies properly. Once these questions have been answered and libraries have a general direction, there are some specific considerations that can be helpful in the service implementation process.

- Budget: How costly are the different service delivery models? What is financially feasible both in the short term and in the long run?
- Staffing: Should text reference service be integrated into existing virtual reference services (e.g. e-mail, IM/chat) for question processing, or run as a separate service point? Is it preferable to maintain the current staffing schedule or to use a different schedule for text reference service? How much extra work is reasonable for librarians to adapt to the new service?

- Usability: What is the acceptable amount of instructions for users? When considering features of the service software (e.g. message organization and display, character limit per response, capability to accommodate collaboration among libraries, capability to send and receive graphics), what is important? What built-in tools of the service software are necessary – character counter, URL shortener, new message alert sound, mechanism to select questions, etc.?

Overall, clear service goals, a detailed picture of the library's wants and needs and a thorough understanding of what each service delivery model offers are key for a library to select an appropriate model to implement and deliver text reference service.

Types of information needs of text reference users

To understand how people use text reference service to meet their information needs, it is important to examine the types of questions they ask. Content analysis was conducted among a random selection of 3103 questions submitted to MIQ in 2010. The unit of analysis was an independent question in a text reference transaction that represented a distinct information need. The coding scheme was developed from Katz's reference typology (Katz, 2001). Table 14.1 presents the distribution of questions across the different categories.

Table 14.1 Breakdown and distribution of question types

Question category	Percentage %
Clearly worded questions	
Local library related	13.3
Non local library related	
Ready reference	69.8
Specific-search	9.2
Personal knowledge related	1.3
About My Info Quest	2.4
Out of service scope	1.5
Unclear questions	2.5

The majority of the questions appeared to be ready reference questions. Ready reference questions are defined as questions that require a single and straightforward answer consisting of a specific and definitive piece of information. This type of question can be broken down in many different ways and across almost all subject areas, for example:

What is the airspeed velocity of an unladen european swallow?
Who won the 1955 superbowl?
What hockey team does sean avery currently play for?
How do i get from the intersection austin st and pecan st to 1634 chestnut st in denton, TX?
Wut does comme d'habitude mean in French?
Will it rain constantly in Peoria illinois today or will it let up?
If you put sickerdoodle cookie dough in the fridige for an hr. will that make them soft cookies when u bake them?

The majority of questions' being ready reference questions is very likely due to the 160-character limit of text messages. Since this type of question seeks a specific piece of information, they are mostly unambiguous and do not involve a comprehensive reference interview. To best answer ready reference questions, librarians should be familiar with reference and information sources on the web and how to effectively search for and locate information within them. A solid knowledge base of web information sources and a network of colleagues from whom to seek help are critical in handling text reference questions. Thus, an easily accessible 'cheat-sheet' containing useful sources on the web and information on how to consult colleagues will be of help to librarians providing text reference service. For instance, a cheat-sheet might include links to library policy/procedures, frequently used reference resources on the web (both general and subject specific), a list of texting lingo and tools such as URL shorteners or character counters that could help librarians to compose their messages concisely.

User perceptions and use of text reference service

In order to develop an empirically grounded understanding of how library users perceive and use text reference service, a survey study was conducted to thoroughly examine its practice from the users' perspective. Both online survey and print questionnaires were made available to the users of seven participating libraries of MIQ. A total of 303 valid responses were received. Among the respondents, 255 (84.2%) never used the text reference service provided by their library, and 48 (15.8%) did.

The non-users of the service were asked about their reasons for not having used the service and their interest in using it in the future. Most of them (59.8%) had not used the service because they were not aware of it; 22.5% explained that they were not texters and therefore the service was not appropriate for them; another 22.5% considered their information needs to be adequately met by assistance from librarians via other means, like telephone, e-mail reference or visiting the reference desk; 15.7% stated that they generally did not ask librarians for help because they were self-sufficient information seekers and were able to use library and other information resources independently; and finally, a modicum of 2.4% offered the following reasons: they were too busy or too forgetful to use the service, issues with their mobile phones such as a broken screen or insufficient texting plan prevented them from using the service and they lacked knowledge of the types of information assistance that were provided by the service.

To measure whether non-users would be interested in using the service in the future, three options were listed – Yes, No and Maybe. More than half (58.2%) of the non-users selected 'Maybe', depending on whether they had the need or not; close to one-quarter (23.7%) opted for 'No'; and a mere 18.1% returned a 'Yes'.

Among the text reference service users, more than half (52.1%) had used the service only once, and the usage frequency of the rest was more or less evenly distributed between 'two to five times' (22.9%) and 'more than five times' (25.0%). They had discovered the service by a number of different means, as shown in Table 14.2.

Table 14.2 How users discovered the availability of text reference service

Means of discovery	Percentage
I found it on the library website.	41.7
I learned about it in the library promotional material/events.	16.7
I heard about it from someone I know.	6.3
I heard about it from a library staff member.	37.5

Users indicated a high level of satisfaction with the service they received. When asked to rate their satisfaction level on a five-point Likert scale (1 = not satisfied at all, 5 = very satisfied), 87.6% selected '4' or '5' as their rating, and only 8.3% considered their service experience as 'not satisfied at all'.

What users liked about text reference service was investigated in order to identify the appealing characteristics of the service. Table 14.3

provides a break-down of the different aspects of the service that attracted users.

Table 14.3 What users liked about text reference service

What users liked about text reference service	Percentage
It's easy to use.	77.1
It's convenient.	66.7
It's fast to get an answer.	52.1
I'm comfortable with it.	47.9
Librarians are a reliable source of information.	4.2
It's cool and fun.	4.2
It is useful when I have an idea of the question I want to ask but cannot form it into a specific Google search.	2.1

The study showed that the majority of library users had not used text reference service. There were three primary reasons for this non-use. First and foremost was a prevailing lack of awareness. In order to make the service known to library users and encourage them to use the service, active marketing campaigns need to be implemented. The study found that users of text reference service had mostly learned about it from browsing the library website or from conversing with a library staff member. Since these were the two most effective means of drawing attention to the service, libraries could devise promotional activities that make use of them.

The second major reason for non-use of text reference service was being a non-texter. Veteran texters are more likely to embrace texting as a means to seek reference assistance, because they are already familiar and comfortable with it. Therefore, it will be helpful if libraries understand who the texting populations are and target promotional efforts towards them.

Thirdly, some non-users are self-sufficient information seekers or are able to have their information needs adequately met by using other reference services, such as telephone, e-mail or visiting the reference desk. To encourage them to try out text reference service, it would be helpful to inform them about what the service has to offer that is not necessarily available via other means.

Although only about 15% of the participants in the study used text reference service, they reported a high level of satisfaction, indicating a positive experience with the service. This is encouraging evidence for libraries that are interested in establishing text reference service. A number of appealing qualities of the service were identified by the study: easy to use, convenient, fast and reliability as a source of information. This knowledge helps libraries to understand what users like about text reference service and what attracts them to use the service, and will help them to devise appropriate strategies to highlight these aspects of the service when promoting the service to their user communities.

Competencies requisite for text reference service

To enhance the professional development of reference librarians and better prepare them for text reference service, it is necessary to identify the competencies requisite for the successful delivery of reference service via texting. Competencies, defined by Griffiths and King (1986) as a combination of knowledge, skill and attitude, are an indispensable component in the advancement of librarianship. The ultimate goal of competency development is to facilitate training and education.

A three-round Delphi study was conducted in a competency identification study. Delphi study is a frequently used methodology in competency identification. It is a technique for gleaning and refining subjective input from a group of people, usually experts, in an attempt to achieve consensus about some aspect of the present or the future. In order to recruit participants, a call for participation was e-mailed to members of MIQ. Another e-mail invitation was sent to the authors of 11 articles about text reference service that had been published in *Reference Services Review* and *The Reference Librarian* in the last four years. Thirty-eight librarians from the expert pool responded and participated in the study.

In the first round, participants were asked to list competencies that they thought to be essential for providing text reference service. A total of 49 competencies were parsed from participants' responses and rank-ordered by frequency of occurrence. In the second round, the rank-ordered list of competencies was sent to participants, who were asked to evaluate each competency's importance to text reference practice on a five-point Likert scale, with 1 being not important at all and 5 being most important. In the third round, a final list of competencies, rank-ordered by the mean value of the importance rating, was sent to participants. They were asked to decide whether or not they agreed with the aggregated group input, and no disagreement was reported.

Table 14.4 presents the ten competencies with the highest importance ratings. As a result of the consensus-building effort, the ten competencies exhibit a substantial degree of agreement among the participants in the Delphi study. As they represent the convergence of the study participants' opinions, they should be carefully converted into learning objectives for education and training programmes.

Table 14.4 Text reference competencies identified by the Delphi study

Competency	Mean value of importance
Ability to compose answers to users' questions concisely, quickly and accurately	4.80
Ability to construct effective search strategies and skilfully search online information sources	4.25
Ability to quickly evaluate information and determine the validity, credibility and authoritativeness of sources	4.05
Knowledge of information resources, especially online information resources	4.05
Ability to interpret users' information needs with limited context in text messages	4.05
Good communication skills, such as maintaining a friendly, respectful, helpful and pleasant tone	3.90
Familiarity with the software/platform used to provide text reference service	3.85
Customer service skills	3.74
Understanding of text reference service policies	3.79
Ability to answer questions politely, intelligently and professionally, even questions that might be judged inappropriate due to language or content	3.74

Three of the competencies in the top ten list – 'Ability to compose answers to users' questions concisely, quickly and accurately', 'Ability to interpret users' information needs with limited context in text messages', and 'Ability to answer questions politely, intelligently and professionally, even questions that might be judged inappropriate due to language or content' – are specific to text reference service because of the unique characteristics of texting as a communication channel, such as the 160-character limit per message and messaging cost. It is not practical to engage in in-depth reference transactions via the exchange of text messages, and hence librarians need to be perceptive and efficient in determining what users are looking for, and to be clear and succinct in providing the information. This can be challenging, as reference librarians are usually expected to be thorough and detail-oriented when assisting library users. Therefore, in text reference training it is just as important to help librarians to understand the shift in expectation and make the proper mind-set adjustment as it is to equip them with requisite knowledge and skills. Topics such as how to use a character counter and URL shortener, as well as how to write concisely without appearing brusque and impatient, are helpful to include in the training programme. It is also necessary to examine transcripts from existing text reference services and to analyse how users frame their questions in text messages. Common messaging patterns will help librarians to better understand users' texting behaviour from the communication perspective, and thus interpret their information needs more effectively and precisely.

Due to the anonymity of texting-based communications, there is occasional inappropriate use of the text reference service, which usually makes librarians uncomfortable. Further, some authentic questions representing genuine information needs can appear improper when the topic of interest is relationships or sex. In order not to dismiss these questions alongside real pranks, librarians need to treat all seemingly inappropriate questions with professionalism and politeness. In text reference training, it is important to communicate this point to librarians and help them to learn to handle professionally and calmly any questions that have uncomfortable language or content.

Conclusion

In conclusion, we hope that the detailed examination of text reference practice will lead to an enriched professional understanding of texting as a reference service method, which will then help libraries to make more informed decisions when considering this service. Our findings indicate that, for library users, text reference service is a suitable means for brief and straightforward information needs; their use of the service is contingent upon their awareness and perception of the service; and they appreciate the ease of use, convenience and speediness of the service; for librarians, it is important to comprehend the affordance of texting and master the skills to interact with users effectively and efficiently in this service environment, such as being able to compose answers to users' questions concisely, quickly and accurately and to interpret users' information needs with limited context in text messages. These findings can be integrated into marketing and promoting the service so as to increase awareness and generate interest, as well as in designing training and educational content to prepare librarians for providing text reference service.

Professional organizations such as the Reference and User Services Association may also wish to consider the findings from this project when updating their behavioural and performance guidelines, so as to incorporate the best practices of text reference. Ultimately, an enhanced understanding of library users in this age of texting will help library and information professionals to gain a more grounded view of how to successfully assist users in their information-seeking by using texting technology, and will help to serve libraries' missions of providing services for the interest, information and enlightenment of all in the user community in a world that is filled with constantly evolving technologies.

References

Griffiths, J. M. and King, D. W. (1986) *New Directions in Library and Information Science Education*, Knowledge Industry Publications.

Katz, W. A. (2001) *Introduction to Reference Work*, McGraw-Hill.

Lippincott, J. K. (2010) Mobile Reference: what are the questions? *The Reference Librarian*, 51 (1), 1–11.

Smith, A. (2011) *Americans and Text Messaging*, http://pewinternet.org/Reports/2011/Cell-Phone-Texting-2011.aspx.

15

M-libraries on the hype cycle: where are we?

Jo Alcock and Pete Dalton

Introduction

This chapter provides a snapshot of current and planned future m-library initiatives and current barriers to their implementation. It concludes with reflections on the current state of m-libraries in relation to the technology hype cycle. Though the focus of the chapter is UK academic libraries, due to the location of the researchers and source of funding, it also draws on examples from outside this area.

JISC-funded mobile library support project

Although mobile technologies and devices have been used in libraries for a number of years, many of the developments have been experimental in nature and not widely reported. In 2010, JISC (the Joint Information Systems Committee) invested in a programme of work to help develop knowledge and skills in the area and share lessons learned with other institutions. Five institutional projects were funded, including investigation of mobile interfaces for accessing academic content (MACON), bibliographic management on mobile devices (m-biblio), sharing specialist library-owned content via mobile devices (PhoneBooth), publicizing information to students through electronic displays and mobile devices (MoPED) and sharing study resources for law students (Mobile LearnMore by LawBore).

In addition to the institutional projects, JISC also funded a community support project aimed to bring together information both from the projects within the programme and from elsewhere and to share this with the wider community. The community support project, managed by Evidence Base at Birmingham City University in collaboration with Owen Stephens Consulting, aimed to 'help support and engage the emerging m-library community by reviewing and synthesising existing research and evidence-based guidance'.

The project (which ran from November 2011 until October 2012) had two main areas of focus: evidence gathering and community building.

Evidence gathering

Under the evidence-gathering part of the project, information and resources were collected from existing projects within libraries utilizing mobile technologies. Information already online was shared by the project through a number of different mechanisms, including social bookmarking services and Twitter, all of which were combined into one RSS feed for people to subscribe to. The tag 'mlibs' was used across services to bring everything together. For information not already online, the project encouraged case studies and guest blog posts to share this information more widely. The lessons learned were particularly valuable for those libraries planning to do similar projects. The resources collected during the project were brought together under a number of different themes to develop a series of Pathways to Best Practice. These included information on the maturity of the area, different options available, examples of implementations and lessons learned. The Pathways to Best Practice covered the following areas:

1. Introducing users to the library
2. Helping on-site users – roving support
3. Helping users within the library – guides and signage
4. Providing access to resources via mobile devices
5. Developing a mobile strategy for the library
6. Managing borrower accounts
7. Bibliographic management
8. Loaning mobile devices
9. Supporting learning, teaching and research
10. Collection management.

Community building

Under the community-building side of the project, an information-sharing event was organized, incorporating a keynote presentation, breakout workshops and discussions and lightning talks. In order to disseminate the information shared at this event more widely, event-amplification methods were used to share tweets and blog posts covering all sessions from the day. The project team also attended, and presented at a number of other events, and hosted workshops on mobile technologies in libraries to share information and encourage discussion and sharing of best practice.

The project blog[1] was used heavily throughout the project as a way to support both evidence gathering and community-building sides of the project; all project materials are hosted or linked to from the blog. An online community website was also set up, but feedback suggested that a mailing list was a preferred option;[2] this continues to be used alongside the blog.

Surveys

To help inform the project, fact-finding surveys were used at the beginning and towards the end of the project. The surveys were hosted on SurveyMonkey and promoted via the project blog, mailing lists and social media. The surveys were open to all, and targeted at those working in libraries and information centres. Although the project was based in the UK, it was hoped to incorporate information from further afield and support the wider library and information science community. The surveys included sections on current m-library initiatives, future m-library initiatives, barriers/challenges to implementation, confidence in implementing, areas where further information is needed and the m-library community. The first survey was open November 2011–January 2012 and had 188 respondents. The second survey was open July–August 2012 and had 138 respondents.

Respondents to both surveys were predominantly from the UK (around 66%), with some from the US, Australia, Canada, Europe, Africa, South America and Asia. The majority of respondents were from academic libraries, with some from public libraries, special libraries, health libraries, government libraries, law libraries and school libraries.

The following sections pull together data from the surveys, in addition to other information discovered during the project, to present an overview of the key trends in current and future m-library initiatives, and barriers to their implementation.

Current m-library initiatives

In both surveys, we asked respondents what their library currently offered with regard to m-library initiatives. As the respondents were self-selecting, it was unsurprising that the majority of respondents already have some sort of m-library offering. In the first survey 63% had; in the second survey 92% had.

Table 15.1 shows the most popular responses (in order) from both surveys:

Table 15.1 Top m-library offerings, as gleaned from surveys

First survey	Second survey
Mobile catalogue	QR codes
Mobile website	Mobile catalogue
QR codes	Mobile website
Supporting use of mobile services/apps	Supporting use of mobile services/apps
Mobile app for the library	Institutional mobile app
Institutional mobile app	Mobile devices to support roving reference/staff demonstrations
Mobile devices to support roving reference/staff demonstrations	Loaning mobile devices
Loaning mobile devices	Mobile app for the library
SMS communication about borrower record	SMS communication about borrower record

As can be seen from the similarities between the two lists, the main differences between the two surveys were the increase in use of QR codes and the institutional mobile app becoming more common than a library mobile app.

Some of the main areas are examined briefly below.

QR codes

Libraries are using QR (Quick Response) codes in a number of different ways. Uses mentioned in the survey responses included to promote mobile resources and e-books, for additional guidance (e.g. link to guidance on how to use equipment), on bookshelf signs, linking to catalogue records and for general promotional activity.

Senate House Library (UK) uses QR codes within its catalogue to help users get information straight to their phone.[3] The QR code embedded in the web catalogue takes users to a mobile version of the record, and was produced in response to the number of users the Library found typing the details or taking a photo from the OPAC with their mobile phone.

One of the concerns about QR codes is whether they are used and whether they provide value or are perceived as just a gimmick. As part of our project, some guides on QR codes in libraries were produced,[4] including examples of uses in libraries, information on how to track QR code usage and general best practice tips.

Mobile catalogue

Many of those who mentioned having a mobile library catalogue were using vendor-supplied modules and specific mobile apps (e.g. AirPAC, BookMyne, Talis Prism). WorldCat, which has its own app and is also supported in a number of other apps, was also mentioned as an option used to provide mobile catalogue functionality.

LibraryAnywhere by LibraryThing was a popular response; this is a useful option for those whose library management system provider does not have mobile functionality, as it works with other systems. Keele University (UK) is one organization using LibraryThing; it has written a case study on this, including information about the requirements, selection, implementation and lessons learned from the experience.[5]

Mobile website

Of those libraries that offered a mobile website, many commented that it had been developed in-house. For some, it was part of a wider institutional mobile website, whilst other libraries had their own library mobile website.

Some mentioned the type of content included on the mobile website. It was commonly a reduced version of the library website, with basic information such as opening hours, location information and contact information, for example:

> We have created a simple mobile website for our library – it does not offer nearly as many options as our full site – but just the most frequently requested information (hours, contact information, Ask Us, etc.).

However, as more people are purchasing mobile devices and mobile web use increases, users are expecting to be able to perform more than the basic functions on the mobile website. Recent research in libraries (for example, a user survey at Oregon State University[6]) has demonstrated that some users expect to be able to perform research searches via the mobile website. Research into how libraries' mobile websites of have changed[7] demonstrates changes in design to support this, including making research links or search boxes more prominent.

Continuing the trend for extending the information available via mobile websites, more libraries are using responsive web design to deliver the same content regardless of the device (with content modules which display in an appropriate way for each device). See Oregon State University[8] for an example of this (resize your browser or access from different devices to see how it adapts).

Supporting use of mobile services/apps

A number of libraries are providing guidance for those of their users who are interested in using mobile apps/websites from other providers, such as publishers. This includes support on enquiry desks, and online guides such as LibGuides with information on available apps/mobile sites and how to use them (see, for example, Newcastle University's Mobile LibGuide[9]). In addition to publisher content, some libraries produce guides for recommended mobile apps such as study guides, productivity tools and document management tools.

Institutional mobile app

Some libraries have functionality integrated into a wider institutional mobile app. Commercial options include CampusM, which provides access to a number of different university systems via its mobile app, including finding available computers on campus and accessing library account details to enable renewals. Many commented on the fact that these apps are compatible with a number of different types of device:

> The University has just introduced a mobile app (compatible with iPhones and with Android devices, BlackBerry to follow) which is student-facing and presents a small selection of services which are all currently available via other web-based services.

Using mobile devices to support roving reference enquiries

Roving reference has been a common development in libraries in recent years, enabling help for users at their point of need, and this has been further supported by mobile technologies such as mobile phones and tablet computers. Libraries in the surveys are using these to deliver support for enquiries around the library, as well as to test and demonstrate library resources.

In the UK, a number of academic libraries are using iPads in and around the library to help users at their point of need. The University of Warwick[10] uses them for enquiries within the Library and also for tracking enquiry statistics (it has developed an app to support this), whilst the University of Huddersfield is using the Roving Librarian brand to help users around the campus by using iPads.[11]

Loaning mobile devices

Some libraries loan mobile devices such as iPads, Kindles and other e-readers to users. For some, this is for use within the library (or on campus), whilst others loan the devices for use anywhere for a period of time. Most libraries loaning devices pre-load them with resources and content, rather than the individual users doing so.

The University of Oxford's Radcliffe Science Library has an iPad loan scheme – guidance is available online[12] and it produces a guide for borrowers.[13]

Future m-library initiatives

When asked whether they were planning future m-library initiatives, the majority of respondents indicated that they were (90.4% in the first survey; 81.8% in the second survey). Many of the initiatives planned are ones that other libraries have already implemented, though some mentioned different ideas, including augmented reality, near-field communication, mobile chat, bibliographic management and support for Bring Your Own Device (BYOD).

For most, there were not yet any plans for specific projects or initiatives as such, just a need to remain updated on mobile technologies and new developments in order to investigate potential uses for the library in future. Developing a mobile strategy was also mentioned by some libraries, either as a separate strategy or incorporating mobile technologies into the broader library strategy.

For those who did have plans for future initiatives, these included the following areas.

Augmented reality

Inspired by the JISC-funded SCARLET project,[14] a number of libraries hope to investigate the use of mobile technologies such as augmented reality to share the material from their special collections to users via mobile devices. The SCARLET+ project[15] is currently testing the original approach in other institutions and aims to share resources to support others planning to do similar.

Near-field communication

Some libraries commented that they hoped to investigate the use of near-field communication to make their services more efficient. Mobile payments are one area facilitated by near-field communication, and something which could be implemented by libraries.

Barriers/challenges to implementation

Respondents were asked to share details of any barriers/challenges to implementation. In the first survey, the lack of technical support was a major barrier, affecting 52% of respondents. Not knowing enough about how to utilize mobile technologies was another key barrier for the respondents of the first survey.

In the second survey, the most common barrier by far was resource constraints, affecting 79% of respondents. Infrastructure/policy constraints and a lack of technical support were also key barriers, affecting 47% and 34%, respectively.

Other barriers included:

- lack of skills
- mobile technologies not being considered an organizational priority
- uncertainty about whether users would want to use mobile technologies
- mobile technologies not being a library/information service priority
- concerns about licensing.

The main barriers are explored in more detail below.

Resource constraints

Resource constraints are a common barrier to much library development, and comments specific to m-library initiatives included issues around a lack of staff time and funding to support new developments:

> Staff have no time for development work.
>
> Financial restraints. I believe we know what we need, but require significant resources to either build something in house, or hire someone to build something for us.

Infrastructure/policy constraints

Barriers regarding infrastructure/policy constraints related to both internal and external contexts and issues at a local and institutional level:

> Currently iPads and other Apple products are not supported by our IT dept.
>
> Institutional Wi-Fi not very reliable.

Conclusion – reflections on the maturity of mobile technologies in libraries

Throughout the project, it was possible to see progress in the area of mobile technologies in libraries. The surveys were not identical (data from open responses in the first survey were used to inform categories for the second), though both surveys explored the same areas. Although details of the respondents to both surveys were not collected, as the same promotional routes were used for each survey it is likely that many of the respondents completed both surveys. As a result, it is reasonable to assume that any changes in results may show progress over time. Even if this is not the case, having two snapshots is a useful way to identify any key trends and look for changes in the area.

Some of the key changes that we noticed through the survey results and through observations during the project included:

- more libraries moving to responsive web design to deliver website content to any device (rather than library mobile apps or specific mobile websites)
- more libraries investing in devices for staff to use for roving enquiries, supporting learning and teaching or to support the green agenda (i.e. less need for paper in meetings)
- more libraries providing training for staff currently unfamiliar with mobile technologies
- more libraries including mobile technologies in their strategy for the future (or having a separate mobile strategy).

When considering where mobile technologies in libraries are in terms of technological adoption, it is useful to consider this in relation to the Gartner Hype Cycle. The Gartner Hype Cycle shows the lifecycle of new technologies and can be used to plot the maturity of adoption of technology (see Figure 15.1 opposite).

Using the data from the survey and during the project, it is estimated that most libraries are between the peak of inflated expectations (buzz around initiatives, success stories but also some less successful experiments with technologies that are not yet established) and the slope of enlightenment (initiatives and technologies becoming more widely understood, many of the initial problems have been solved). The project ran a workshop at the 4th International M-Libraries Conference in September 2012, in which participants were required to indicate where they perceived their own institutional developments to be on the Gartner Hype Cycle. Generally, the responses supported this view, with many participants placing their markers between these points. There were, however, some who felt that their library was still at the technology trigger point, and it is clear that there are still barriers to be

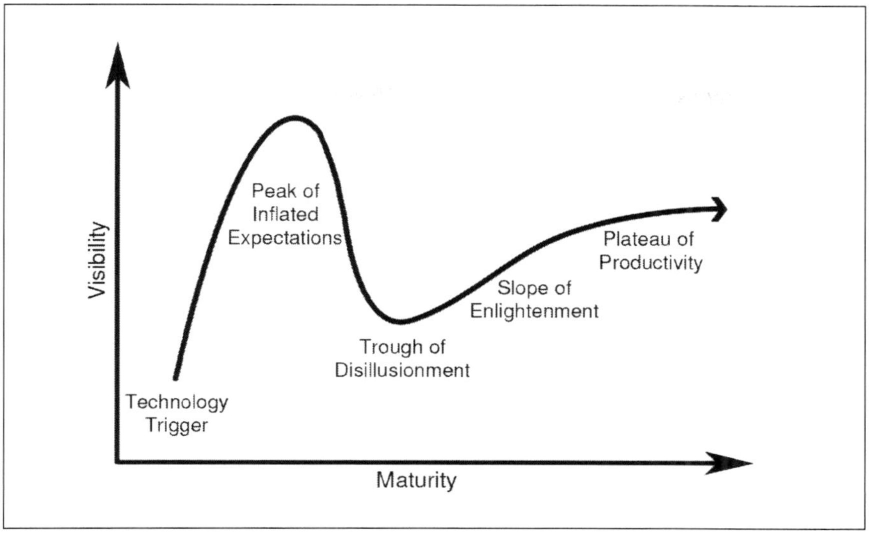

Figure 15.1 The Gartner Hype Cycle[16]

addressed before we reach the stage of mainstream adoption where all libraries are at the plateau of productivity.

Research during the course of the project has shown progress in the area of mobile technologies in libraries, but it is clear that there is still a way to go. It is hoped that resources such as the M-Libraries conferences and this book will help to share best practice and reduce duplication of effort, helping libraries to learn from each other and keep up with technological developments.

Notes

1 http://mlibraries.jiscinvolve.org/wp/.
2 m-libraries-group@jiscmail.ac.uk.
3 http://mlibraries.jiscinvolve.org/wp/case-studies/qr-codes-at-senate-house-library-a-case-study/.
4 http://mlibraries.jiscinvolve.org/wp/tag/qrcodetips/.
5 http://mlibraries.jiscinvolve.org/wp/case-studies/libraryanywhere-at-keele-university-a-case-study/.
6 http://storify.com/joeyanne/lita-mcig-virtual-meeting.

7 www.slideshare.net/bohyunkim/its-time-to-look-at-your-librarys-mobile-website-again.
8 http://oregonstate.edu/is/home#.
9 http://libguides.ncl.ac.uk/mobile.
10 http://wrap.warwick.ac.uk/35367/.
11 http://eprints.hud.ac.uk/14523/.
12 www.bodleian.ox.ac.uk/science/services/e-readers/ipad.
13 http://ox.libguides.com/ipad2.
14 www.jisc.ac.uk/whatwedo/programmes/elearning/ltig/scarlet.aspx.
15 www.jisc.ac.uk/whatwedo/programmes/elearning/embeddingbenefits2012/scarlet.aspx.
16 www.gartner.com/technology/research/methodologies/hype-cycle.jsp.

16

Responsive web design for your library website

Matthew Borg

Introduction

Most written pieces and presentations on mobile technology inundate the reader or listener with a plethora of statistics detailing the past, current or future state of technology. It is astonishing how often these statistics back up the position that the speaker or author has adopted! No matter what source one relies on for statistics, one inescapable fact is that the number of people accessing the internet from a mobile device is on the rise. A late 2012 report from the Pew Research Center suggests that the number has risen from 25% in 2008 to 56% in 2012 (Duggan, Rainie, Zickuhr, 2012).

Higher education institutions are no strangers to developing and deploying apps for mobile devices. Research suggests that as of January 2013, more than 80% of UK higher education institutions either have an institutional app, or are planning to introduce one in the next 12 months (CAPITA, 2013).

This article describes the choices made by the Virtual Learning Centre library team at Sheffield Hallam University (SHU) when deciding how to present the Library interfaces to students and faculty. Part of the Virtual Learning Centre's role is to focus on how library systems can be designed and implemented to facilitate and enhance an enjoyable experience for the user.

How are our users viewing content?

When SHU was preparing to launch its first app for mobile devices in summer 2010, the Virtual Learning Centre team was asked to contribute a mobile version of the library website so that students could undertake simple tasks such as searching for and renewing books. As mentioned, data on mobile are everywhere. But for me, the most important data are about how people are looking at the content I am creating. How many people are looking at our library website on a mobile device? How are they using our site? What device specifically are they using to view it? What are they looking at when they are there and what are they saying about it afterwards? We use standard analytical software on our Library website (Google Analytics, Reinvigorate) and are able to pull this data out. For example, for the period January–October 2012, the number of times an iPhone was used to access the library website was up 301% on the previous year. During the same period, visits via iPads were up 987%.

Mobilizing with SHUgo

Mobile interfaces to websites have traditionally been cut-down versions of full sites. 'When people begin to think about building mobile applications, a common first thought is to build a website that is optimized for mobile browsing' (Clark, 2012). For a long time, web teams designed sites to present information based on desktop displays, and mobile design was 'an afterthought' (Wroblewski, 2011).

The natural thinking of the development team was therefore to create a mobile, stripped-down version of the Library website. We had ensured that we were able to extract management information. We already knew, for example, how many users were accessing the website via a mobile device. We were also able to ascertain how 'popular' the links on the front page of the site were, using the aforementioned analytical tools. We came up with a prototype mobile version. A cut-down version of the site, with just a few key links to our resource discovery tool (Summon, from Serials Solutions), the Library catalogue and renewals. By this point, the institutional app had a name – SHUgo. As well as integrating the mobile version into SHUgo, we also used some simple code on the main Library website that could detect whether a mobile device was being used and would redirect users appropriately.

The site worked well and integrated neatly with SHUgo. We started to do some user testing of the mobile version of the website. Once we started to show the mobile version alongside the desktop version, the limitations of the optimized site began to show.

Design wise, there was a massive difference between the two sites. It was a jarring experience for the user to move from the desktop website onto the mobile version. In the first round of user testing a common request was for a link that would take the user to the desktop version. Some users wanted tools that we had removed, based on our 'expert' intuition. This made me question the practice of automatically creating a cut-down, optimized version for mobile use.

Time to abandon the printed page

The general consensus seems to be that there are two main reasons for designing stripped-down, mobile-optimized versions of websites. Firstly, that we should be designing for smaller screen real estate, and secondly, that people generally use their mobile devices to access data when they are on the move.

It is time to challenge these positions. An excellent starting-point for us was a seminal article on designing for the web: John Allsop's 'A Dao of Web Design'. In this article, Allsop suggests that 'It's time to throw out the rituals of the printed page, and to engage the medium of the web and its own nature' (Allsop, 2000).

The nature of the web is fluid. There are numerous screen sizes, and it is a Herculean task to design specific interfaces for all of these sizes. Historically, web design has been subjected to the limitations that print design has experienced. You will find multiple print design treatises that mention 'above and below the fold'. This refers to newspapers, where important stories were generally placed at the top of the front page so that when the paper was folded, this material was still visible. This paper-based design trend moved into web design. The portion of the screen that is visible before the user scrolls down was considered sacrosanct. But the web is not paper based. It should not be limited by the same restrictions. As Thereisnopagefold.com (2013) suggests, 'Welcome to the World Wide Web, an interactive medium in which screen resolution statistics are trivial, browser viewports are variable and scrolling behaviour is a standard.'

Rather than challenge ourselves to present information to fit specific dimensions measured in pixels, we should adopt the fluid dimensions that the web allows us to experience. Are we trying to aim for some sort of perversion of Ranganathan's third law: Every device its own design? Instead, implementing a responsive website allows our design to be device agnostic.

Additionally, the school of thought that suggests that people use their mobile devices to access information on the go is also out-dated. We just need to look at a range of research from the last couple of years. A recent survey undertaken by Google and IPSOS tells us that 93% of smartphone users use their device at home, 72% at work, 54% in a coffee shop and 32% in a library (Think with Google, 2011). This is not the behaviour of people that solely use their mobile internet connections when they are in a rush and on the move. Even the smallest room in the house gets a look-in – 75% of Americans have used their mobile phones in the bathroom (11mark, 2012)! A most startling statistic is that almost a fifth of phone owners do most of their online browsing on their phone rather than on a computer or other device (Smith, 2012).

Mobile first

So we tried not to use our expert intuition and instead listened to what our usability testing was telling us. We developed a new version of the Library website using responsive web design as a key tenet of our development. The initial designs were created for mobile devices and then expanded to see what they would look like on tablet and desktop screens. Designing mobile first enabled us to focus on a first-class user experience which was not dependent on the 'correct' device being used.

There are essentially three elements to designing a responsive website. Media queries are used to detect the type of device accessing the website. They can detect the capabilities of the device, such as the dimensions of the screen that is being used. The relevant CSS style is then delivered so that the website can adapt to the size of the web browser being used. The design of the site is done using a fluid grid. In this, all the page elements are set to relative units, rather than absolute. For example, we set the font size to be a percentage, rather than a fixed pixel weight. That way, the individual elements, including fonts, can adapt and resize according to the web browser. Finally, we use flexible images by sizing them according to relative screen size. For example, we deliver the main image on the Library website as 100% of the column it is in, rather than, say, 400 pixels.

Conclusion

We need to be aware of our users, and design for them. As Ethan Marcotte suggests: 'If we're willing to research the needs of our users, and apply those ingredients carefully, then responsive web design is a powerful approach indeed' (2011). The key thing here is that we are building and maintaining a single, responsive website that can be served as a stand-alone interface, or integrated into the institutional app. The best of both worlds.

References

11mark (2012) *IT in the Toilet*, www.11mark.com/IT-in-the-Toilet.

Allsop, J. (2000) 'A Dao of Web Design', *A List Apart*, http://alistapart.com/article/dao.

CAPITA (2013) 4 out of 5 HE Institutions Embracing Mobile Technology to Improve Student Experience, *CAPITA Further and Higher Education*, www.capita-fhe.co.uk/news/Pages/4outof5HEinstitutionsEmbracingMobileTechnologytoImproveStudentExperience.aspx.

Clark, J. (2012) *Building Mobile Library Applications*, The Tech Set 12, Facet Publishing.

Duggan, M., Rainie, L. and Zickuhr, K. (2012) *Mobile Connections to Libraries*, Pew Internet and American Life Project, http://libraries.pewinternet.org/2012/12/31/mobile-connections-to-libraries

Marcotte, E. (2011) *Responsive Web Design*, A Book Apart.

Smith, A. (2012) *Cell Internet Use 2012*, Pew Internet and American Life Project, http://pewinternet.org/reports/2012/Cell-Internet-Use-2012.aspx.

Thereisnopagefold.com (2013) www.thereisnopagefold.com.

Think with Google (2011) *The Mobile Movement – Understanding Smartphone Users*, www.thinkwithgoogle.com/insights/library/studies/the-mobile-movement/.

Wroblewski, L. (2011) *Mobile First*, A Book Apart.

Part 3

Implementation

17

A new method of training users: Polimedia video for iOS and Android devices

Angels Carles-Pomar, Ana Castellano López and Fernando Guerrero Rebollo

Introduction

There is no doubt that mobile technology has changed our lives in many ways – from getting organized at work, to how we communicate and even our study or play habits. Libraries are not the exception that proves the rule. In recent years we have seen how libraries had made great efforts – and also huge advances – to leverage technology and its application in order to improve and expand their services. Traditionally, borrowing books has been one of most important and valued services for library users. After the loan service, one of the main library functions is to provide access to information, and to disseminate information. Therefore most libraries invest time and expertise in information literacy.

For several years, Universitat Autònoma de Barcelona Library Services (UAB SdB) has been organizing an introductory face-to-face course for all 'first time' students to explain how to access library information resources. This includes instruction in how to consult the bibliographic database and how to locate any kind of document. The course, called 'Vine a conèixer la teva biblioteca' (Come and get to know your library) is common to all UAB libraries and has had very satisfactory results.

The course was designed to help students to develop skills and capabilities needed not only for their studies but also throughout their careers. This type of training cannot be forced; it is only when students realize that they lack the ability to find the literature that is relevant to their needs that they will take the time to learn. User education is an investment for the future. The users acquire the skills to find relevant information in the library and take advantage of its bibliographic resources, thus making knowledge accessible to those who need it. Although none of this is new, the fact is that users' preferences are increasingly moving towards flexible attendance rather than fixed courses.

Thanks to the increasing use of mobile devices, users have access to information from anywhere and anytime, and as often as they want, allowing much more learning to take place than in traditional teaching models. Thus, ICTs help to improve the learning process and reflection on dynamic concepts that are usually difficult to understand in lectures. Overall, libraries are lagging behind in terms of understanding and using information technologies – their impact on work habits, their capabilities and limitations, the use of e-learning resources, and an understanding of how students use mobile devices in their daily lives.

Technology in the library – a little history

Generally speaking, the introduction of technology into libraries occurred during the 1980s. In the case of UAB libraries, it was in 1990. Bibliographic catalogues and circulation desks were the first to benefit. Today no one can imagine a library without searchable, automated catalogues. In recent years, technology has also reached user training courses, with all kind of tutorials – from how to browse the catalogue or how to borrow a book, to laptops, lecterns, USB sticks, personal digital assistants, or any type of technological device for use in the library. Nowadays, this is not a new question for us. But there is another question. *How* could the technology be much more useful and interesting for students? What is the key to success? In implementing a new way to teach and learn in the library, could it be a tool that teachers could use to improve theoretical as well as practical lessons?

How to improve the research skill of the user

The main objective of the Medical Library is to collaborate with other services of the UAB Medical School – such as the Learning Resources

Centre and Computer Services – to improve the informational skills of users. One way or another, this goal has always been present in the library's functions, only the format is changing: we must move with the times, providing the users with easier and quicker ways to learn how to access relevant and precise information. In order to achieve our goal, we needed technical support from the Learning Resources Centre and Computer Services.

Background

In 2002, the UAB Medical School formulated a plan to address the challenges of creating the new EEES (European Area for Higher Education). The aim of this plan was to promote and increase the use of multimedia learning by encouraging the use of computerization and audiovisual media in teaching. One of its consequences was the creation of the Learning Resources Centre. The main aims of the plan included the introduction of ICTs (hardware and software) into the faculty and supporting the development of teaching materials in order to make use of these technologies. Since 2002, when the Learning Resources Centre was created, the production of this kind of teaching material has increased.

Some workgroups and teachers from the UAB Medical School have been involved in projects to improve and promote the introduction of new technologies as part of the teaching methodology. One such workgroup was in contact with colleagues from the Universitat Politècnica de València (UPV). UPV has developed a specific tool for online courses, which is known as Polimedia.

The first contact between the UAB Medical School and UPV began in 2005. UAB was interested in using Polimedia to create virtual lessons as a supplement to traditional teaching. In October 2008, UPV and UAB Medical School signed an agreement authorizing and allowing the UAB Medical School Learning Resources Centre to use Polimedia. Accordingly, a recording studio was created at the UAB Medical School. Furthermore, since 2005, the Learning Resources Centre and the Medical Library have collaborated on several projects. The first project was to design a library tutorial for users on searching for bibliographic information.

Nowadays, any catalogue, databases and all search engines are easier to use and friendly. In addition, students and even teachers are familiar with querying them, and many users are digital natives. Nevertheless, there are users who are not part of the 'net generation', and they may feel uncomfortable asking for help. In addition, in UAB Medical Library there are other types of information resources that require some explanation in order to be used efficiently. For example, the bibliographic manager Refworks; or how to access the major subscription databases like Scopus or Web of Knowledge; and also the free and open databases, such as PubMed, the most important dataset in medical sciences, produced by the National Library of Medicine in the United States. In order to create tutorials for these types of resources, we needed a teaching tool, and that was why we decided to introduce Polimedia.

What is Polimedia?

Polimedia is a brand-name tool, designed and developed at UPV, as noted above. It enables the production of multimedia content to support teaching and is a production system designed for the creation of quality training materials to strengthen and complement traditional lessons, but without being a substitute for these. It allows the teacher to create a range of virtual materials, and provides students the opportunity to learn and accumulate knowledge, anywhere and anytime, and to review the videos at their own convenience.

The success of Polimedia lies in its originality – a key feature that differentiates it from other tutorial tools. The Polimedia product is a multimedia resource combining, simultaneously and in the same space, two video recordings. While one video shows the speaker's explanation, the other one displays the content that is being projected, e.g. a slideshow, a software application, internet searches and so on. Polimedia's versatility allows the creation of videos with all kind of variations. For example, you can make a tutorial using and combining two videos, with or without the teacher speaking in the image, or just create a demo video presentation.

Teachers like the system because of its many advantages. Lessons are pre-recorded, which means that the explanation needs to be given only once, without any repetition. If the student does not understand the subject matter, they can play the recording again, as many times as necessary. Meanwhile, the teacher can focus on other tasks to reinforce learning. Further, Polimedia allows much more learning to take place than in traditional teaching models. It helps to improve the learning process and promotes better understanding of dynamic concepts that are usually difficult to understand by just reading.

Polimedia also has some advantages for students, allowing them to have control over their studies, since they can decide when, where and how often they access the learning materials. They are responsible for their own workflow, which is beneficial because each student is different. Because of its open and flexible nature, Polimedia means that each student can decide from where to access the information. They can watch the videos several times, and more than one user at a time can watch the same video, as all the videos are stored on the same *streaming* server, enabling simultaneous multi-user viewing.

Are all these great features of Polimedia adaptable to library services? Of course they are! Polimedia is a good solution for library user training, such as showing how the self-checkout works. We decided to implement Polimedia video recordings as a new service in the UAB Medical Library. Collaboration between the Learning Resources Centre and the Medical Library greatly facilitates the production of videos for this new service, because sometimes it is difficult for librarians to manage and resolve technical problems. The Learning Resources Centre handles the recording process and the Medical Library makes the videos available to the library users.

Materials and methods

The Learning Resources Centre has a recording studio of 30 m^2 with lined walls to improve the acoustics and soundproofing. The set where the speaker stands (Figure 17.1 on the next page) – directly in front of a white background – is illuminated by three cold light sources.

The recording process operates as follows: using a gyroscopic mouse (long range), the speaker can control the slideshow of the presentation, which is also recorded on video. A multimedia technician records two videos simultaneously: one of the slideshow and another of the speaker in action. The final product is a video in which the speaker appears on the screen and the slideshow is displayed to the right of him/her (i.e. on the left side of the screen). The effect is thus that of a teacher standing

Figure 17.1 Polimedia set

next to a blackboard (see Figure 17.3, page 163). The two images, speaker and slideshow, are synchronized; editing and subsequent publication are almost immediate.

The technicians recommend that the video should not exceed ten minutes in length. Ideally, the duration will be between five and ten minutes. Other recommendations relate to the speaker's dress, such as avoiding wearing transparent clothes, bright colours, large prints or thin stripes. Dark colours are definitely the best choice, due to the contrast with the white background. It is also advisable not to wear long necklaces or bracelets, so as to avoid noise interference.

The staff of the Learning Resources Centre comprises three specialized multimedia technicians, each with a specific role, such as recording, complementary programming tasks and promoting Polimedia videos to faculty members.

The Learning Resources Centre has the following recording equipment (Figure 17.2):

- two compact camcorders
- one digital audio table
- one wireless system 'wireless' microphone
- three cold spotlights
- one digital video mixer

Figure 17.2 Polimedia equipment

- one interactive audio compressor
- one handset study
- two stereo speakers
- one tablet PC
- one gyroscopic mouse
- two personal computers.

The videos are hosted on a streaming video server that allows several people to view the same recording simultaneously. Additionally, the multimedia technicians have developed a web platform whereby users can search for videos by subject, course name, author, title or date. This platform is hosted on a separate server, in a three-layer architecture. The web platform is programmed in AJAX (Asynchronous JavaScript), a group of inter-related web development techniques used on the client-side to create asynchronous web applications, written in PHP, and powered by two databases created in MySQL, an open source relational database management system. Each of the three layers has a specific function: the first, a private zone for data management, contains information about the author or speaker, the name of the institution or department, the video title, video duration/length (in minutes), subject and metadata. The second layer, which is also a private zone and is restricted to the administrator, is devoted to statistical data and contains information on searches such as the number of hits, most-visited videos, number of request by countries, number of visits per speaker, etc. The third layer is the public area, where users, via a web page, can access the content hosted on the streaming video server and perform queries about the videos by news items, subject and most-visited records. The online public database can be consulted on Polimedia website (http://polimedia.uab.cat/).

Polimedia videos in the UAB Medical Library

UAB Medical Library has two types of users. The first type are teachers, researchers, investigators, postdoc, PhD and others with special requirements. The other type are students. Both groups need to find the best possible information related to their research areas or to their curricula.

The Library must attend all users, but is more sensitive to the former group. There is a reason for this: not only are the undergraduate students younger, but they also belong to the internet generation and therefore they are not afraid of technology, since technology is usually part of their lives. Students do not feel embarrassed or awkward when they have to ask a question. This may be a common trait among students of this generation, regardless of the library in which they are regular users. The Medical Library holds tutorials about library services that are also common to other libraries in the UAB – catalogue searching, locating books, creating lists of records or preferences files, request for documents and book renewals being among the most relevant. And it does this despite the fact that, increasingly, the bibliographic tools are more friendly and easier to use.

Polimedia tutorials are appropriate for explaining complicated topics, because one of the advantages of Polimedia is the ability to view the same video as many times as necessary. It is often necessary to view a video more than once in order to correctly understand some complicated concepts that are difficult to assimilate. In some instances, the focus is on showing the use of bibliographic tools: how to perform a search, how to save the results of a search, or how to get the results of a search on mobile devices. However, our main interest is in supporting researchers. Terms such as H index, IF (Impact Factor), SJR (SCimago Journal Rank), SNIP (Source Normalized Impact per Paper), besides knowing how to find them, need no longer be a mystery. Knowing the number of publications and number of citations per year, or what is the best paper published in a particular scientific subject, etc., has become a requirement for researchers.

Other relevant information may be to know the percentage of non-cited items, or to compare the results of different scientific reviews in up to ten titles. This requires specific training, and in our opinion the best way to carry this out is by using Polimedia videos (Figure 17.3).

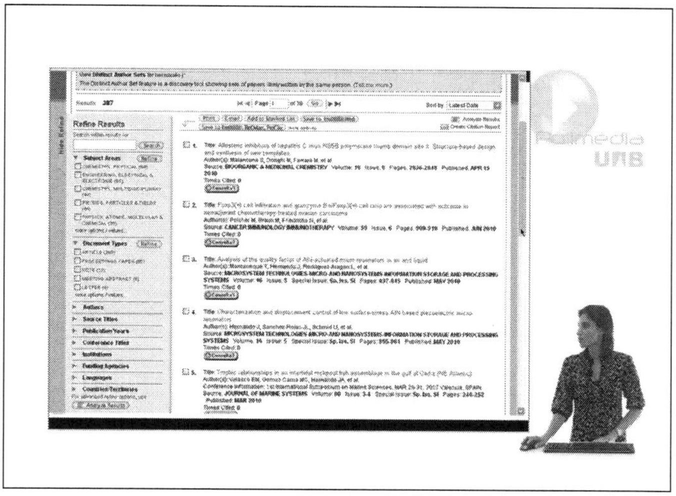

Figure 17.3 Polimedia video (reproduced with permission)

What students think of Polimedia

The students participated in the experience of recording videos, giving oral presentations to other students. A survey was conducted by a teacher, who introduced the Polimedia tool and asked the students their opinions of it.

Of 19 responses collected from students, 14 considered the experience a positive one (73.68%), and 4 very positive (21.05%). Only three students did not want to try it again, and all but one recommended it. Students who had watched the videos several times recommended them to their peers, friends and family. For 12 of the students (63.15%), recording a video of this type did not mean extra preparation effort when it comes to show it to their teammates. Fifteen students (79%) prefer restricted view than open access. In our opinion, students' work would have a greater impact with open access.

Conclusions

We believe that Polimedia is a very good tool with which to support teaching and learning in a self-directed learning environment. It is also excellent for training in bibliographic search skills, delivered by specific training tutorials.

Technology is changing fast, but not all users are able to learn quickly. For young students, especially those who use ICTs, Polimedia is an ideal solution. The recordings provide an opportunity to improve their skills and knowledge without having to attend classroom courses at scheduled times. Older users often feel uneasy asking the librarian for help at the information desk. These users have a strong interest in information literacy but have difficulty in assimilating some of the concepts. Polimedia is an ideal solution in these cases and for practising repetitive tasks. As a tool, it stands out for its flexibility and privacy. It also provides a new way to keep in touch with the teacher; and users can access the training videos database everywhere, anytime and as often as necessary, without having to ask for explanations, and without showing vulnerability or fear in the face of technology.

We are convinced that this kind of training format is a suitable means for training Medical Library users in bibliographic search skills.

18

Mobile phone technology in academic library services: a public university student's perceptions and paradigm

Anwarul Islam

Introduction

At Dhaka University, changes in society and technology have resulted in changes in the students' use of technology and in their views and attitudes towards using libraries. In this context, new means of communication are necessary in order to keep in touch with library users. New technologies have made communication and access to information very convenient for users, both from the comfort of their own home or office and from wherever they are while on the move with their mobile phones or personal digital assistants (Karim, Darus and Hussin, 2005). The use of mobile technology is increasing, especially in the Asian countries such as Bangladesh. Bangladesh is one of the few countries in the world that can guarantee that every one of its residents can get a mobile phone signal – no matter where they are in the country, with its population of over 150 million. City Cell was the first mobile phone company in Bangladesh, starting operation in 1996, and the total number of mobile phone subscribers had reached 94.714 million by the end of July 2012 (BTRC, 2012). At present, there are six operators providing mobile phone services in Bangladesh (Table 18.1 on the next page).

Mobile phone applications in education and library services

The widespread use of mobile phone technologies, as compared to the

Table 18.1 Mobile phone providers and subscribers in Bangladesh	
Operators (Six)	**Subscribers (million)**
Grameen Phone Ltd (GP)	39.556
Orascom Telecom Bangladesh Ltd (Banglalink)	25.622
Robi Axiata Ltd (Robi)	19.652
Airtel Bangladesh Ltd (Airtel)	6.806
Pacific Bangladesh Telecom Ltd (Citycell)	1.391
Teletalk Bangladesh Ltd (Teletalk)	1.685
Total	**94.741**
Source: www.btrc.gov.bd/	

use of personal computers, can be clearly seen across all walks of life in Bangladesh. Their use is also widespread among students in institutions of higher learning. This may provide broad opportunities for institutions and businesses alike to apply the technology for commercial as well as for educational purposes. Mayer (2002) has proposed several steps that institutions can follow to provide SMS services. These steps involve forming a collection of mobile phone numbers, setting up a centralized SMS service centre and sending group SMS via a network. Information services delivered by SMS can be reading lists, lectures, academic schedules, exams, homework reminders and so on. Many higher education institutions in Bangladesh have been providing various mobile-phone based information services. Among them are Dhaka University (DU), Jagannath University (JU), Jahangirnagar University (JU2), Bangladesh University of Engineering and Technology (BUET) and Rajshahi University (RU). A summary of the services provided is shown in Table 18.2. As can be seen from the list, none of the services provided is library related.

With many institutions aiming to position themselves at the forefront of tertiary education, it is inevitable that mobile solutions will be the best means for providing convenient service on the modern campus. This can be done by extending the service delivery channels from PC to mobile devices via SMS, and through WAP over GSM, GPRS and 3G technologies. For instance, library users can now renew books and pay fines for overdue loans without entering the library, so long as they are carrying their mobile phones. In the case of library services, wireless technology has the potential to offer many new

Table 18.2 Universities and mobile-based information services

Services	DU	JU2	JU2	BAU	BUET	RU
Exam results	•	•	•	•	•	•
Admission process	•	•	•	•	•	•
Student intake information	–	•	•	–	–	–
Registration	•	•	•	–	–	–
Class schedules	–	–	–	–	–	–
Date and venue of exams	–	–	–	–	–	–
Helpdesk	–	–	–	–	–	–
Complaints	–	–	–	–	–	–
Campaigns	–	–	–	–	–	–
News	–	–	–	–	–	–

Source: University websites and University Grants Committee manual

possibilities for accessing information from online catalogues, online databases, the internet and virtual libraries. Among the wireless technologies are digital cellular technologies, packet radio and wireless local area networks. Today, the convergence of mobile phones and the internet through the WAP standard presents libraries with a real opportunity to deploy wireless phone technology to manage their operations for the following services:

- checking records of books borrowed
- receiving alerts about overdue books
- receiving alerts about outstanding fines
- receiving reminders to return library items that will be due soon
- renewing library items
- reference enquiry services
- receiving text alerts about new resources on the library website
- getting information from the library OPAC/database
- contacting the librarian for help.

Some examples of mobile phone use in library services are shown in Figures 18.1 and 18.2 on the following page.

Currently, libraries are looking for ways to communicate effectively with their users. Although the applications are limited to notification, no libraries in Bangladesh are making an effort to use wireless services for such purposes, but such services will inevitably come in the future.

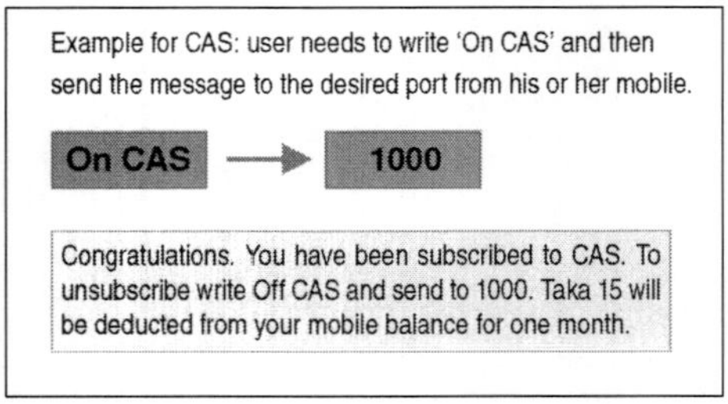

Figure 18.1 Subscription to current awareness service via mobile phone

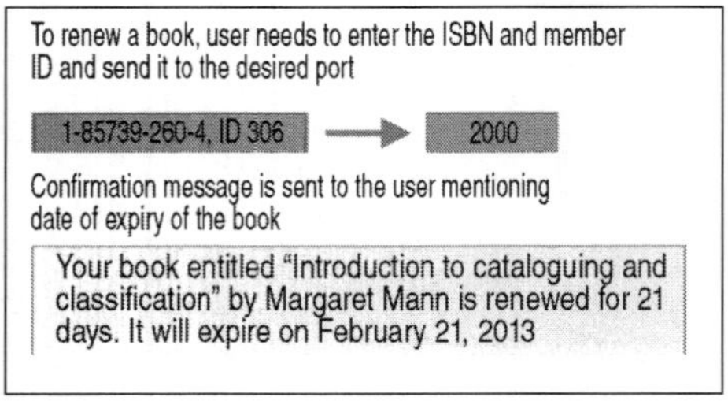

Figure 18.2 Book renewal service via mobile phone

When the time comes, both academic institutions and telecommunication companies need to be well equipped with knowledge and understanding of the concepts that underpin provision of library and information services and of what the users need.

Objectives and methodology

This study seeks to explore the utilization of mobile phone services in the educational environment, to explore the nature of mobile phone use among university students and investigate university students' perceptions of the use of mobile phones in library and information services. The study was conducted with students of Dhaka University, the oldest and largest public university in Bangladesh. The data for the study were gathered from undergraduate and graduate students in the faculties of Arts and of Social Science. The Faculty of Arts has 16 departments and Social Science has 11 departments. The study used a survey questionnaire for data collection. In early June 2012, students of ISLM (Information Science and Library Management) and MCJ (Mass Communication and Journalism) were contacted in their scheduled classes and given an explanation of the study. To maximize the response rate, students were advised that their responses would be completely anonymous and the data would be used only for the purposes of the study. A printed version of the questionnaire was distributed in the class. Data collection began in July 2012 and continued through the last week of August 2012. Of the 290 questionnaires distributed, 237 were returned, giving a response rate of 81.72%. A few responses that either were incomplete or contained skipped questions were dropped from the analysis.

Data analysis

The questionnaire was designed to elicit information on respondents' demographic, their experience in using mobile phones, their average expenditure, their awareness of various available mobile phone services, their patterns of library use, their perception of the types of library services that should be offered via mobile phones, their willingness to participate if those services were offered and their views on the level of payment that would be acceptable in order to cover the cost incurred. The respondents were also asked qualitative questions about their views on wireless applications in library services.

Demographic information

The respondents were almost evenly distributed among the two academic programmes: ISLM (51%) and MCJ (48%). However, the distribution of the respondents was uneven in terms of year of study, gender and nationality. The majority were in Masters classes (37%), female (54%) and had 6–7 years' experience of mobile phone use (19%) (Figure 18.3 and Table 18.3 on the next page).

Experience of using a mobile phone

It was very encouraging that most of the respondents were mobile phone users and just 9.3% (22) students had not used a mobile phone. Their experience as mobile phone users varied from 1 year to more than 8 years.

The majority of students (45, or 19%) had been using a mobile phone for 6–7 years, followed by 44 (18.6%) for 5–6 years and only 3 (1.3%) for less than one year. Only 9 (3.8%) students had been using a mobile phone for more than 8 years. GrameenPhone has the largest number of subscribers in Bangladesh, at more than 39 million, but the majority of the students in the study (29%) used Banglalink (which is very close to GrameenPhone, at 27.4%). It is apparent from the data that most of the students (87.8%) were using a pre-paid plan and most of them (37.1%) spent more than 300 BDT per month (1 Tk. (taka) = US$85). The result might perhaps have been different if the respondents had been academic faculty members, as the respondents did not have a reliable income.

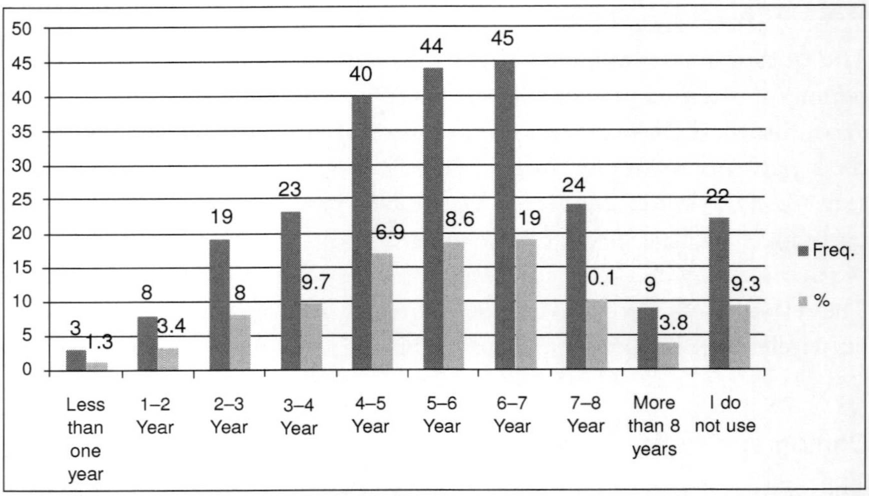

Figure 18.3 Characteristics of mobile phone use

Table 18.3 Respondent demographics

Variable	Type	Frequency	Percentage
Department	ISLM	121	51.1
	MCJ	116	48.9
Gender	Male	109	46
	Female	128	54
Study level	First year	32	13.5
	Second year	32	13.5
	Third year	34	14.3
	Fourth year	51	21.3
	MA	88	37.1
Age	15–18 years	6	2.5
	18–20 years	40	16.9
	20–23 years	106	44.7
	More than 23	85	35.9
Experience	Less than one year	3	1.3
	1–2 years	8	3.4
	2–3 years	19	8
	3–4 years	23	9.7
	4–5 years	40	16.9
	5–6 years	44	18.6
	6–7 years	45	19
	7–8 years	24	10.1
	More than 8 years	9	3.8
	I do not use	22	9.3

Use and awareness of mobile phone services

The students' use of wireless services available via their mobile phones was also analysed (Table 18.4).

Table 18.4 Distribution of respondents according to application services used (multiple)

Application services	Frequency	Percentage
SMS	198	83.5
GPRS	78	32.9
MMS	70	29.5
WAP	67	28.3

The figures are low for WAP services. This may be due to lack of awareness among the respondents, who were receiving less exposure to progress in ICT development.

Table 18.5 provides information on respondents' use and awareness of various information services available via SMS, GPRS and WAP. The results show that all information services offered by the telecommunication service providers were, at some point, being used by the students. The highest usage was of news (60.3%), followed by e-mail (40.1%), entertainment (38.8%) and sports (36.7%). The results also indicate that most of the respondents were aware of the information services even though they were not using them.

Table 18.5 Usage and awareness of information services (multiple)

Information services	Frequency	%
News	143	60.3
Sports	87	36.7
Banking	44	18.6
Ticketing	36	15.2
Entertainments	92	38.8
Games	65	27.4
E-mail	95	40.1
Others	63	26.6

Library use

Information about library use was acquired through questions about the frequency of library visits and the purposes of the visits. The results show that a majority of the respondents (46%) visited the library only when necessary. On the other hand, a reasonably high percentage of respondents visited the library a few days a week (25.3%). Overall, most of the respondents had visited the library at some point. Not much difference was observed between the two academic programmes in terms of the students' library visiting habits. Information about the purpose of library visits is shown in Tables 18.6 and 18.7.

Table 18.6 Purpose of library visits

Purpose of library visit	Never	Seldom	Sometimes	Often	Always	Mean	SD
Research for assignment	3.8	8	44.3	25.3	18.6	3.47	1.01
Borrow or return books	24.1	18.1	35	15.6	7.2	2.64	1.21
Read for test or exam	1.7	13.5	27	32.5	25.3	3.66	1.05
Discuss with group study	10.5	27.4	29.1	23.6	9.3	2.94	1.14
Use the internet facility	24.5	22.8	29.5	15.6	7.2	2.58	1.22
Know current global news	20.3	22.8	26.6	20.7	9.7	2.77	1.26
Audiovisual materials	44.3	19.8	19	14.3	2.1	2.1	1.18
Ask questions of the librarian	21.1	23.2	31.6	18.1	5.9	2.65	1.17
Others	11.8	30	29.1	22.4	6.8	2.82	1.11

The information was collected using a frequency scale of 1 (=never) to 5 (=always). The results are displayed in terms of mean distribution and the standard deviation. The closer the mean of the total score to 5, the higher the tendency of the respondents to visit the library for the respective purpose. The results show that among the most popular purposes for library visits are reading for exams, research for assignments, group study discussion and borrowing or returning books.

Table 18.7 Borrowing books

Methods of borrowing	Never	Seldom	Sometimes	Often	Always	Mean	SD
Using OPAC	40.9	13.1	20.7	17.3	8	2.38	1.36
Library website	27.4	15.6	26.2	15.6	15.2	2.76	1.4
Through circulation desk	11.8	15.6	24.9	28.7	18.6	3.27	1.27
Information desk	14.8	22.8	26.2	25.7	10.5	2.95	1.23
Others	19	14.3	36.7	20.3	9.3	2.86	1.21

Use of mobile phones in academic library services

In general, students were asked what they thought of the use of mobile technology in libraries. Most of the students (78.1%) thought that it is possible to use mobile technology in libraries, followed by 14.3% partially possible. 'I do not know' was 5.5% and 'no' was 2.1%. The study also investigated the methods which students used to check their borrowing status, using a frequency scale of 1 (=never) to 5 (=always). This information was elicited because of the high possibility of mobile phone applications focusing on this area of library service.

The results indicate that most of the respondents were utilizing the circulation desk (mean=3.27) for this purpose. The results suggest that students were generally less interested in the other methods, as most ratings fell below 3, i.e. the average score. The mean scores below 3 may indicate that respondents did not favour those services. Another possibility might be lack of awareness about the library website services, and feeling shy or intimidated to ask the librarian at the service counters.

Perceptions about mobile phone applications in library services

This section provides findings on the respondents' perceptions in relation to the potential use of mobile phones for various library services. Respondents were asked about the perceived usefulness of the proposed services, should they be offered by the university. Preferences for the various library services were measured on a Likert scale of 1 (=not at all useful) to 5 (=very useful) (Table 18.8).

A mean score higher than 3 indicates that the respondents were in favour of the service. The results show that most of the respondents

Table 18.8 Respondents' perceptions of mobile phone applications in library services

Perceived mobile phone application (features)	1	2	3	4	5	Mean	SD
Renewing of library items	2	3	13	49	34	4.1	0.9
Getting information from the library	0	1.3	11	43	45	4.3	0.7
OPAC/database	–	2.5	16	45	37	4.16	0.8
Checking records of book borrowed	–	0.4	12	56	32	4.19	0.7
Alerts for overdue books	–	0.4	13	49	38	4.24	0.7
Information on outstanding fines	–	0.4	14	54	32	4.19	0.7
Reminders to return library items	–	0.4	14	39	46	4.32	0.7
Reference enquiry services	–	–	10	42	48	4.38	0.7
Current Awareness Service (CAS)	–	–	7.6	39	54	4.46	0.6
Help text message (text a librarian)	–	–	19	42	38	4.19	0.7
Read e-book	–	0.4	18	45	37	4.19	0.7
Opening hours, working days	–	0.8	13	41	46	4.32	0.7
List of new resources by subject	2	0.8	5.5	44	47	4.34	0.8

were in favour of the provision of current awareness services via their mobile handsets. Other proposed services also received a high mean score, indicating that most of the services proposed were considered to be useful by the respondents. These results were further enhanced by the fact that 85% of respondents indicated that they would subscribe to the library services if they were offered. The remaining respondents were probably reluctant for reasons such as affordability and accessibility.

Finally, general comments were also gathered using an open-ended question. Respondents were asked to state in their own words their opinions about the proposed wireless library services. The following are examples of the responses:

- It is important because sometimes I do not have much time to go to the Library because I will be busy with the assignments and projects in the computer lab.
- It is useful for those who stay off campus!
- Users may access library services after working hours.
- It is one of the effective ways for the Library services. It makes the Library convenient for us in various ways.
- It is a good idea because wireless application is a new trend that gives more flexibility.

Findings and future work

It is apparent from the study that the application of mobile phone technology is not new in many universities in Bangladesh. These universities are already providing services to their users through SMS. On the other hand, the provision of library-related services is still considered new, and no universities in Bangladesh are offering mobile-phone based library services. But the findings of the study were that the respondents were willing to utilize library services through their mobile phone devices. The widespread use of mobile phones among students may have led to positive opinions and perceptions of mobile applications for library-related services. The lack of such applications is perhaps due to the many constraints of wireless-networked applications and devices, the complexity of library services and operations and the lack of research on database and online retrieval processes in a wireless environment. However, university libraries in Bangladesh could start to deliver passive information services, such as library operating hours and basic circulation-related information. Services such as borrowing and renewal status should be a priority for Bangladeshi academic library services. Such services could also be utilized by other types of libraries, such as public libraries, corporate information service centres and the national library networks, for better access. Such an application was also viewed as beneficial by the respondents, perhaps due to the fact that they visited the library less frequently, and only when necessary. The low level of visits should not be seen by librarians as a threat, but instead as an opportunity to be more visible to users. Wireless application services, if offered, would allow libraries to reach a wider range of user groups, giving libraries more opportunities for better promotion.

Conclusion

The results of the study have provided necessary information on what users really want. Thus, the study has perhaps provided a signal for libraries to begin participating in these kinds of services. Telecommunication providers need to produce a good study investigating the appropriateness of library functions for delivery in the smaller interface environment, like the other mobile phone services already offered. The results of such a study should be used in the actual implementation of the services.

This study was not without its limitations. Although it was assumed that variability in the sample was low in relation to the variables under study, greater differences might be observed with a bigger sample size and with other categories of library users, such as academic staff. Another similar study may be needed to confirm the validity of the results.

References

BTRC (Bangladesh Telecommunication Regulatory Commission) (2012) *Mobile Phone Subscribers in Bangladesh*, www.btrc.gov.bd/.

Karim, N. S. A., Darus, S. H. and Hussin, R. (2005) Mobile Phone Applications in Academic Library Services: a students' feedback survey, *Campus Wide Information Systems*, **23** (1), 35–51.

Mayer, I. (2002) *Using Text Messaging to Improve Student Organization and Motivation*, http://ferl.becta.org.uk/display.cfm?resID ¼ 2762&printable ¼ 1.

19

Moving beyond the counter: mobile library support and the use of tablet PCs at Leeds University Library

Peter Kilroy

Introduction

This chapter documents an attempt by Leeds University Library to trial a mobile 'Library Support' concept, in which iPads and Android tablets are being used to bring the functionality of the counter directly to Library users at the point of need. This represents a convergence of two important processes that are becoming increasingly common across higher education libraries in the UK and beyond: the shift to self-service systems and burgeoning mobile strategies. The chapter offers a broad overview of the project, focusing on its background and technical implementation, as well as outlining some of the logistical challenges and possible future developments. It concludes with a provisional list of dos and don'ts for implementing tablet-orientated roving reference systems.

Background

The Library Support project began just in 2011, when Leeds University Library decided to adopt a full self-service system comprising a set of 3M self-service units and one 3M book sorter. (The machines were in use prior to the Library Support project, and have now become the main means of issuing and returning books.) This meant shifting the emphasis away from some of the basic functions performed at the counter, such as issuing and returning books, and towards a more advanced and all-round enquiry service that we branded as Library Support. We used bright lanyards, shoulder bags and display-screen posters to make the point visible (other visibility options that were considered, e.g. bright T-shirts, having proved unpopular with staff). We wanted the service to be available to our users at the point of need, which meant not only redefining the role of customer services staff, but more fundamentally – and more ambitiously – also redefining or rethinking the role of the University Library itself: from a static counter to a more dynamic team of roving staff, and from relatively passive to more pre-emptive and proactive (Brown, Sulz and Pow, 2011).

Choosing a mobile device

Of course, the obvious question for such roving staff was this: how do we bring some of the counter functions to staff and students at the point of need? Our initial solution was to think of using smartphones, but it quickly became apparent that these would be too limiting in terms of screen size (Widdows, 2011). We wanted to be able to show students the process of our enquiries, not just tell them the results. However, we also envisioned roving staff doing more than just answering enquiries (e.g. checking the multi-function devices, finding lost items, tidying the shelves, ensuring that students were not eating or making excessive noise, etc.) and therefore didn't want anything too large either.

As a result, we initially settled on the seven-inch form factor, which at that time meant Samsung's Galaxy Tab Android tablet. We also put together a very simple Android app for performing searches and finding information, using what was then Google's App Inventor service (http://appinventor.mit.edu). However, we ran into some compatibility issues between Android and our Eduroam Wi-Fi system, and so we decided to keep the Galaxy Tabs for the sites where it worked and, as iOS users had not encountered any Wi-Fi problems, we ordered a batch of second-generation iPads for the other sites. (The iPads have subsequently become our default tablets because (a) staff find them more convenient and pleasurable to use, (b) they are more compatible with our Wi-Fi system and (c) it is easier to develop for one platform.) We also purchased a set of 'Library Support' shoulder bags in order to free up our hands during roving (Widdows, 2011).

Although the situation has since changed (see, for example, Sharman and Walsh, 2012; Williamson, 2012), preparatory research on the use of tablet devices within libraries revealed two important trends: first, that they were often used by librarians for back-room rather than front-end functions; and second, that they were often used with home page links to mobile library catalogues or other websites or apps, rather than with a custom interface (e.g. Brown, Sulz and Pow, 2011; Lotts and Graves, 2011; May, 2011). Our plan, in contrast, was to use iPads and Galaxy Tabs primarily for front-end functions, and to devise a custom interface rather than using home page links.

However, as we were now working across platforms, screen sizes and aspect ratios, we decided to minimize the problems of fragmentation by developing one HTML5 web app, which could be tweaked for different devices and easily updated. This 'device agnosticism' (PC Mag, http://bit.ly/VoHGBy) also had the advantage that we could use the style templates laid down by the recently redesigned Library website (http://library.leeds.ac.uk), which we trimmed and optimized for tablet screens. Our guiding ethos was that any prospective web app had to be as simple as possible, with a minimum of visual distractions and a range of different functions tailored to the needs roving staff.

Technical implementation

The web app is composed of a top banner (linking to the Library home page), two navigation bars (top and bottom), one search bar and a series of large buttons, each serving a clear function. The top navigation bar offers a series of lesser-used but permanent links: 'Routes' and 'Campusmap' for directions, 'Floorplans' for navigation and orientation and the 'Library Guide' for an introduction to our services. Where appropriate, these links launch the relevant iOS or Android apps (an Apple ID for each device being required on iOS). For example, 'Campusmap' launches Apple or Google Maps; and our various social media feeds – RSS, Facebook, Twitter and YouTube – launch their respective apps. We are trying to future-proof by focusing on social media, whilst giving roving staff a chance to catch up on the latest events and activities within the Library.

The bottom navigation bar offers a series of lesser-used but term-specific links. Presently, these are 'Skills@Library', 'Special Collections' and 'Researcher@Library'. In this case, we are bearing induction in mind, with a number of skills and research resources, including videos explaining basic library functions. The principal aim is to use such resources and videos for meeting and greeting new students at the turnstiles and for queue-busting – explaining how to use our services to those waiting, in order to speed up the process. At other times of the year we can change these links as necessary (for example, offering exam skills videos during exam time, etc.).

Central functions

However, the main emphasis of the app is on the central block of buttons. This is something we change regularly, depending on feedback (sent via the feedback button on the bottom navigation bar, which allows staff to send a pre-addressed e-mail), but the idea is that the most used functions go here, and this is a good indication of the kinds of enquiries we get in order of priority.

The first row covers the fundamentals: searching the catalogue and checking Library records. The 'Search@Library' bar is our quick, smart-search function, designed to work without the immediate need to specify filters. Searches can be performed by text, voice and – on Android devices – by using the tablet's camera as a barcode reader via the Barcodescanner Keyboard app (http://bit.ly/WsMVuX). The catalogue recognizes ISBN numbers, and so publishers' barcodes can be scanned.

The principal idea here is a scenario where, for example, a student comes to us with a high-demand or short-loan item and asks whether we have any other copies. Rather than typing a search and trying to find the right edition, we can scan the ISBN number straight into the catalogue. If the enquiry is more in depth than 'Do you have …' or 'Where is X book?' we can do an advanced search, and we can also check Library records if required. Again, students can either sign in directly or – on Android devices – we can scan their cards using the tablet's camera.

The central row of buttons is reserved for more general information, covering the different Library sites, opening hours and IT information. This last is particularly useful, as we can give students live status updates about Wi-Fi outages or server problems, etc., both of which are common enquiries.

The bottom row of buttons is reserved for back-room functions: contacting the desk and recording enquiry statistics. The 'Contact Desk' button opens a pre-addressed e-mail which, when sent, creates a pop-up notification on the counter PC using Google Talk (www.google.com/talk). Desk staff can use the instant messenger function of Google Talk to create a drop-down notification on the tablet. (The reason for this workaround is the need to develop a communication system that (a) works from within a browser and (b) works across platforms.) In most cases, we use this system to recall roving staff when it is busy at the counter, but it can also be used to ask staff to find missing items or to bring books back to the counter for a waiting student, etc.

Enquiry stats and feedback

Finally, roving staff can record an enquiry by using the 'Enquiry Stats' drop-down box (choosing from 'Item Enquiry', 'A/C Enquiry', 'IT Enquiry', 'Directions' or 'Other'). This sends enquiry data to an automatically populated spreadsheet using Google's cloud-based Drive service (https://drive.google.com). The spreadsheet is set up to provide an enquiry summary as well as an hourly, daily and monthly breakdown, and this can be viewed live by clicking on the 'Stats' link on the bottom navigation bar. It also automatically creates a series of infographic charts to visualize the data and to determine any interesting trends (e.g. a peak of certain enquiries at certain times of the day or certain times of the year, etc.). We can then feed this information back into the development of the app by changing the relative position of the most important and most used functions.

Logistical challenges

As you might imagine, implementing a project of this sort has not been without its logistical challenges, and it may be worth outlining a few of these briefly. The first relates to a wide skills gaps between those familiar and unfamiliar with such devices, and by 'unfamiliar' I don't mean a straightforward skills gap to be filled by conventional training, but a genuine apprehension in the face of a whole new model of touchscreen computer interaction. The extent of this apprehension took us by surprise and we had to work through it slowly in small and informal training sessions to get people used to the idea. This parallels similar challenges faced by other university libraries attempting to implement tablet use for roving (e.g. May, 2011; Widdows, 2011; Williamson, 2012). However, the clear lesson from our trial was that the larger screen of the iPad and the relative simplicity of the iOS operating system made it a more popular choice amongst staff (Brown, Sulz and Pow, 2011, 10–13).

The second challenge relates to a concern about removing the familiarity of the counter and asking staff to be more proactive. This was also approached by offering informal and small-group training, in this case provided by external trainers from a psychology/counselling background. This training explored scenarios and challenges (e.g. the potential conflict between roving support and behaviour management), as well as possible techniques (e.g. how to achieve a balance between firmness and fairness, or assertiveness and flexibility, etc.).

The final challenge relates to the perceptions of the students themselves. It is often assumed that university students are at the cutting edge of technological developments and are open to – and excited by – all things new. Our experiences have not contradicted this idea per se, but have directly paralleled the skills gap that we noticed amongst staff. Whilst some students have embraced the prospect of a mobile Library Support service, others have found the idea alien to their perceptions of what a library is, with the counter still often seen as the knowledge and information hub, particularly in the smaller libraries. As a result, we feel that it may be some time before a mobile Library Support service is accepted as a mainstream norm. However, it is partly with these kinds of logistical challenges in mind that we are looking to future developments.

Future developments

The first potential development relates to the issue of efficiency: how do we more efficiently bring together roving staff wanting to help and students requiring help without depending solely on observational skills and serendipity (Brown, Sulz and Pow, 2011, 8)? The obvious answer, which we are working on, is to allow students to communicate directly with roving staff, and we have a number of possible solutions in mind.

The first involves using 'short codes' (i.e. the short text numbers often used for services like bus or train times). In our case, the idea would be to divide the Library into a series of locations that the students could text to the service, thereby triggering an e-mail to roving staff. This has the advantage of being inclusive (ownership of basic feature phones remains high [Arthur, 2011]), and it is a discreet way to ask for help or to report problems, but it would most likely involve subscribing to a commercial texting service.

Another option would be to use QR codes programmed with a pre-addressed and pre-titled e-mail (e.g. 'I am in location X. Can I have some assistance please?'). As this is pre-written, students would simply need to scan the code and press send. However, our preferred option at present is to do something similar via instant messaging, allowing for more flexible, two-way communication. These options have the advantage of being cheap and easy to implement and use but, despite the fact that smartphone ownership is on the rise versus feature phones (Arthur, 2011), this may still be too exclusive for some libraries.

Whatever the choice, some form of communication between students and Library staff would likely increase efficiency, although, as with the project more generally, it may take time for such a system to be accepted as a mainstream part of our Library services.

Finally, related to this is the possibility of producing a customizable student version of the app, allowing students to search the catalogue, check their record and contact a member of the mobile Library Support team, etc. However, that would involve moving to a more responsive or multi-device design. The favoured alternative at present is to use and develop native apps for more general student and staff use, including Blackboard Mobile Learn (http://bit.ly/WfvGkd, http://bit.ly/M9JFTI) and the University of Leeds' own iOS and Android apps (http://bit.ly/11Z4wBt, http://bit.ly/WCuODL).

Conclusion

By way of conclusion, we have compiled a provisional list of potential dos and don'ts for implementing tablet-oriented roving reference systems:

1 Consider the size and scope of the library (is roving appropriate/necessary?).

2 Consider surveying students about whether a roving service would be beneficial.
3 Consult with staff at an early stage to ensure that roving does not appear to be imposed.
4 Consider whether you have sufficient staff numbers to provide a consistent service.
5 If you are planning to use a tablet PC, consider which screen size is likely to be the most appropriate for you (larger screens – particularly iPads – are easier to use and more popular with staff, but may make other tasks more difficult).
6 Think carefully about (i) what functions you think you might need to perform (e.g. checking the catalogue, checking library records, contacting the desk, recording stats, etc.); (ii) whether you need a custom interface to perform such functions; (iii) whether your services will be compatible with your operating system; and (iv) whether or not all of this can be done in-house.
7 Consider how you might make roving staff more visible (e.g. bright shoulder bags or lanyards, but bear in mind that clothing options may well prove unpopular with staff).
8 Consider how you might publicize the roving service (e.g. bookmarks, pamphlets, display screens or via the website).
9 Consider what other libraries are doing (e.g., York and Leeds university libraries arranged reciprocal visits to share experiences and ideas about, amongst other things, the use of iPads for roving – indeed, many of these dos and don'ts reflect that exchange).
10 Don't underestimate the level of training required to get some staff used to new technologies (focus on small groups and peer training).
11 Don't underestimate the level of training required to get some staff used to new ways of working away from the familiarity of the counter (having other tasks helps).
12 Don't assume what students' needs are without consulting them.
13 Don't overestimate students' familiarity with new technologies.
14 Don't underestimate students' familiarity with new technologies.
15 Don't underestimate the psychological barriers to asking for help ('This might be a stupid question, but ...' [Brown, Sulz and Pow, 2011, 3]).
16 Don't wait for students to ask questions; anticipate their needs.

Although this implementation project is still at a relatively early stage of development, our hope is that the trial will offer a glimpse of how university libraries might move beyond the functional limitations of the feature phone/smartphone and the mobility limitations of the desktop, towards a more innovative and mobile library support service that does not begin and end at the counter.

References

Arthur, C. (2011) Smartphones Head to 'Tipping Point', *Guardian Online* (11 July), www.guardian.co.uk/technology/2011/jul/11/ios-android-blackberry-smartphone-data.

Brown, C., Sulz, D. and Pow, V. (2011) *Roving Reference with iPads: a study of the use of iPads as technological support and service assessment*, International Federation of Library Associations, http://conference.ifla.org/past/ifla77/197-brown-en.pdf.

Lotts, M. and Graves, S. (2011) Using the iPad for Reference Services: librarians go mobile, *College and Research Library News*, **72** (4), 217–20.

May, F. (2011) Roving Reference, iPad Style, *The Idaho Librarian*, http://theidaholibrarian.wordpress.com/2011/11/23/roving-reference-ipad-style.

Sharman, A. and Walsh, A. (2012) Roving Librarian at a Mid-sized, UK-based University, *Library Technology Reports*, **48** (8), 28–34.

Widdows, K. (2011) *Mobile Technology for Mobile Staff: roving enquiry support at Warwick University*, University of Warwick Institutional Repository, http://wrap.warwick.ac.uk/35367/1/WRAP_Widdows_Mobile_Technology_for_Mobile_Staff_-_final_April_2011.pdf.

Williamson, R. (2012) *Using iPads for Roving Enquiry Services*, M-Libraries, www.m-libraries.org/williamson.

20

Creating a university-wide mobile app: the mStir experience

Andrew Wilson

Introduction

Confusion and pressure in the marketplace have motivated many higher education institutions, especially since about 2010, to create centralized university-wide mobile apps. University marketing and promotion departments' rallying call 'We need an app!' has initiated development without any real research or understanding of who would use the app and whether it is needed. The general 'buzz' around apps has led universities to consider them to represent an advantage over their competitors. However, the benefits that such apps could bring were not then clear. Universities are large, multi-threaded, diverse communities with many different departments operating as separate entities, and this complexity muddied decisions about what to include in an app, leading to apps that were either lightweight on functionality (with many external links to the institution's website), or overburdened with unusable functionality. As is often the case with new technology, time has resolved many of these issues. Development of university-wide apps has progressed beyond these initial problems, with the emergence of stable platforms, both at the mobile device operating system level and at the development level. This chapter, using the University of Stirling's app, mStir, as an example, will offer a starting-point for institutions considering developing a university-wide app, covering the common development platforms, some key decisions made along the way and the anticipated problems.

Why

The University of Stirling started development of mStir, a cross-platform web app, in 2010. We began development at this early stage because we wanted to experiment with the new technology and discover what frameworks and toolkits were currently available. We also wanted to investigate how quickly an app could be developed using these frameworks. Since our development began, the mobile device market has matured (for both hardware and software) and universities, including Stirling, have learned from early mistakes, finally understanding the desires of their target audience. Therefore the present time (2012) is an excellent time for institutions to consider developing an app.

Over the past few years, the tumultuous market conditions, caused by operating system manufacturers competing for the largest share of a fast-growing mobile device market, have led to time- and resource-limited developers at universities struggling to target the key devices for their specific audiences. Over time, the market has stabilized, with some operating systems coming to the forefront. This stabilization has allowed developers to work on university apps confidently, targeting devices in the knowledge that the majority of their mobile-device-owning audience will be able to use the app. The website Mobile Statistics[1] gathers information from a variety of reputable industry research companies, such as Canalys and Gartner, to summarize the current market for 2010 and 2011 (2012 statistics are unavailable) (Figure 20.1). The following can be gleaned from their analysis:

- Android sales continue to rise to over 50% market share.
- iOS sales rise over time, but steadily.
- BlackBerry sales remain flat and stagnant.
- Microsoft Phone 7 sales remain flat.[2]

The app audience

So, the technology has improved, but this is meaningless unless you know your audience. The audiences for a university-wide mobile app are usually internal users: staff and students. Students often drive mobile app development forward. Younger students are confident with mobile devices and appear to desire a university app, but, when asked,

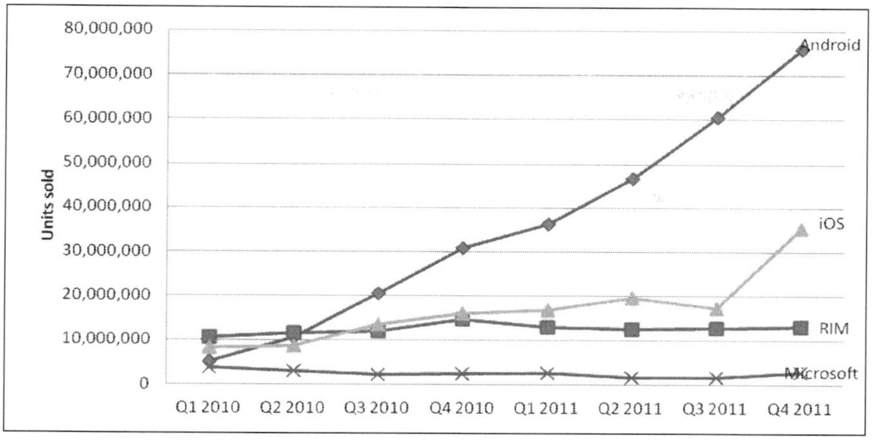

Figure 20.1 Worldwide mobile device sales per quarter, 2010–11 (based on quarterly device sale graphs on mobilestatistics.com)

they are not entirely sure what functionality the app should have: they just want an app. This vague desire has previously undermined the market for university apps. Indeed, the University of Stirling's Information Services IS Satisfaction Survey results in 2010 highlighted that one of the most requested features of a university app would be e-mail (Davey, 2010), something that the University has offered on mobile devices for many years already, with full instructions for all major operating systems.[3] The lack of clarity regarding users' needs is something the developer and the University will continue to negotiate.

Clarity is a significant strategic issue for university app development: clarity of the market, clarity of demand, and clarity of a development pathway within the large and dispersed university structure. Developing an app, however, can provide a valuable overview of this structure, its key systems and departments. Developers can discover, combine and utilize the data that is available from these dispersed systems for the app, often with surprising and useful results.

With the app in mind, a developer might acquire data from the systems that can then be mashed with other data from completely unrelated systems to produce unexpected functionality. Developing an app is often the easy part, accessing the required data for the app to consume is difficult. Once this data has been accessed, it has scope beyond the app, such as broadcasting bus schedules on digital displays and server status feeds for websites.

How

Creating an app for a university is about negotiation and compromise: negotiating access to key data and systems across the university, and achieving a compromise between this available data and the expectations of the app's target audience. A familiar example of this is providing personalized timetables for students in an app. Students want this, but the way the data is held at a university (room bookings and timetabling are often on separate systems), means it may not be possible to achieve. User expectations of the functionality of an app will be high, but the data that a developer has access to may fail to fulfil this expectation. Data access was a problem at Stirling. The University of Stirling is a relatively small, campus-based institution and, as such, liaising with other teams to access data was perhaps easier than for some larger universities; however, accessing some data was not possible, due to time constraints, lack of, or costly Application Program Interfaces (APIs) and vendor-locked data (access to which would breach any service agreement).

Once data has been obtained, what functionality should an app have? At Stirling we used student survey data and discovered the key devices used by the target audience (students) and the type of services the audience wanted to use on an app. The service, although in beta format in 2010, was launched with much of the key functionality the students wanted. The developers attempted to match the quick wins of the easier-to-access data at the University with the desires of the students. Cross-departmental working groups were created to oversee and discuss the functionality of the app, and whether the data that were required could be garnered quickly enough.

Surveys can help, but there are also some useful guidelines:

- App functionality must be task or experience driven for the user, rather than overwhelmed with information. Common examples of this task-driven process are using terminology such as 'I want to …', for example, 'I want to check my essay grades', or 'I want to find out when the next bus home is'.
- Information should be in small snippets, for access on the move, rather than replicate information from the main university website. *A mobile version of the university website is not a university mobile app.*
- Links to external resources, such as websites, should be kept to the

minimum, with provision in place for a mobile version of the resource.

After the data accessible from the university has been summarized and matched to the expectations of the key audience, two vital development decisions are needed: whether to develop internally or use a vendor, and whether to create a native app or a web app.

Vendor vs in-house

The choice between creating a service in-house or recruiting an external company is a decision that is made many times and across different departments at universities, and software development is simply another service to consider.

For app development, there are two major types of vendor-based solutions: off-the-shelf products such as OMbiel's CampusM and Blackboard Mobile Central that slot into your university's infrastructure, or bespoke apps developed by software houses. These options are very different: the bespoke product can be more flexible and less homogenous; however, if platforms or operating systems change future development may be difficult, while the off-the-shelf solution is usually more future proof.

Resource questions

The university's resources may dictate whether you use a vendor (for a bespoke app or not); the university may simply not have the development expertise in-house to work with multiple operating systems, but it may not be able to afford to use a vendor. Many universities also specialize in specific programming languages to standardize much of the development platforms and hardware, and so changing to a different language for one product may not be possible. Vendors should have the expertise to develop the app in the native languages required for the mobile devices that the university wishes to support. If the product is to be developed in-house, the staff resource implications can be significant, and are often higher than those for using a vendor. If a vendor is chosen, however, the university is still required to commit considerable time to making the correct data available in the correct format for the vendor to feed into the app.

Problems with data remain whether the university chooses a vendor-based or an in-house app. To get around some of these problems, at Stirling we used a combination of in-house and vendor solutions. The University bought in vendor-based products to work with the in-house app, accessing particularly difficult data. For example, the University encountered problems in interfacing fully with the Library catalogue, as expensive APIs were needed. To combat this, a vendor-based third-party solution, Library Anywhere, was purchased to enable full Library catalogue integration – mStir then acted as a hub linking through to this third-party application.

User interface

The user interface (UI) is also an important factor when considering whether to take an off-the-shelf product. Adopting a product that already exists will limit the possibilities to change the product's look and feel. The product may be similar to other apps from other universities and seem rather generic. At a time when universities are in competition, every detail may count in attracting their key audiences, whether new overseas students or external businesses wishing to use university venues for conferences. If the product is developed in-house, the UI can be tailored to the exact needs of the university, creating a unique, fresh interface.

Support

When something goes wrong with an app, who do you turn to for support? Generally, vendor support is slower than in-house; it is quicker to pick up the phone to call a server team within the university than to wait for a vendor to deal with a support call. Conversely, the support offered by vendors can be incredibly useful to the university in the long run. When operating systems are updated for mobile devices, the apps for them need to be checked to ensure that they are compatible, and, if they are not, an update needs to be rolled out. With an off-the-shelf vendor solution, this is usually taken care of without any intervention from the university, whereas if the product were developed in-house, developers would need to check the apps every time a new update was released.

Working in-house enables a developer to experiment with different frameworks, UIs and features to truly get the experience they want for the university app. A vendor-produced app will often dictate a path which, although convenient for the university, may not be best for the app and the services it interfaces with. Using a vendor may also lock the university in to a certain avenue for a period of time, restricting the ability to change later. This can be mitigated, however, by ensuring that the data provided to the vendor app is of a recognized standard, using formats such as XML, and APIs or web services that can easily be transferred to other apps or services in the future.

Native app vs web app

The next decision to make is whether to program a native app for each mobile device supported by the university or to make a web app. This decision is influenced by the choice of vendor or in-house. A vendor may offer only certain options, or even take the decision out of the hands of the university by supplying native apps for all major mobile-device operating systems. Alternatively, if the university wishes to develop in-house, it may choose a web app, as there is no need to develop in multiple languages. At Stirling, the vendor/in-house decision was considered after development had begun. Developing and experimentation had been progressing in-house for many months before vendors were invited to the University. This enabled the app working group to compare the experimental app developed in-house with the functionality offered by vendors. It was found that although the vendors offered many advantages, such as native cross-platform apps and extensibility for future market changes, the amount of work required to format the University's data for the vendor apps was too great, especially as data had already been formatted differently for the in-house app. It was decided that, for the time being, the development of the app would continue in-house.

A native app is an application developed in the native language of the operating system on which it runs. This application code is compiled and run natively by the operating system. The app is usually obtained through a store, which handles any updates and maintenance. A web app, however, is a mobile website, developed using modern mobile standards and frameworks to imitate the look and feel of a native app, offering much of the same functionality, but able to run across many different mobile device operating systems. The terms 'native' and 'web' app are becoming more ambiguous with the advent of frameworks such as Phonegap, which uses one language and distributes the app across many different operating systems as a native app.

During the experimental stage at Stirling, it became clear that developing a web app would be the best course of action for the University, mainly because the lead developers were web developers. This decision was also influenced by the rapid release of many university-specific open source web app frameworks during 2010, including UCLA Mobile Web,[4] the Molly Project,[5] Kurago,[6] MIT Mobile Web[7] and Mobile Web OSP.[8] These frameworks differed in their functionality, some offering 'out-of-the-box' elements such as library access and VLE (virtual learning environment) integration, while other frameworks focused on the look and feel of the app, allowing developers to create their own bespoke functionality. After experimentation, the University settled with Mobile Web OSP, a fork of the MIT Mobile Web framework that offered a combination of built-in functionality with a user interface that was reasonably future proof. It is worth noting that development of Mobile Web OSP has now ceased, but, as it was future proof, updating the app for new devices and operating systems is straightforward.

User experience

Both native apps and web apps have advantages for the developer, both visually and functionally, but user expectation is a consideration in this decision too. Web apps, although able to mimic native apps, will not have the exact UI of a native app. The web app at the University of Stirling mimicked a native app, but although students can open the web app, and add the app as an icon on their home screen which, when opened, looks and functions like a native app, verbal feedback suggested that students still wanted a 'real app'. Native apps will, if coded correctly, have an excellent, fast, native UI that the user will already have used on their device, helping the overall user experience of the app. The speed of the app will also usually be faster natively because compiled native binaries are faster to run than interpreted code such as JavaScript. However, in late 2012 Sencha, a proponent of HTML5, developed 'Fastbook', an HTML5 mobile version of Facebook reputed to run much faster than the natively coded official Facebook app.[9] This highlights that with the advent of mobile-specific libraries such as JQuery Mobile, this speed differential is becoming less of a factor.

One of the most influential limiting factors in the development of a web app over a native app is device feature access. Native apps can access all the features built into the mobile device, such as GPS, cameras and accelerometer. Web apps using HTML5 can access only GPS at the moment, although future HTML5 developments will include more feature access. Another crucial device feature that may not be accessed is offline storage. Many university apps consist mainly of time-sensitive information, such as bus times or the latest course module news; however, other app features such as maps and some static content do not often change. Having the ability to use offline storage enables users to access some features of the app without an internet connection. Traditionally, offline storage was the domain of the native app, but the advent of HTML5 Web Storage[10] enables persistent data even if the user closes the browser, allowing web apps to handle offline storage.

Web apps usually have the benefits of a long life. They are websites that can run, without any intervention, for long periods of time on a web server. If an update is required, the update is applied to the web app, and then all a user need do to update is refresh the page. Native apps must be updated via the store for that particular operating system and must go through that store's testing procedure. Native app stores' testing can add costs to a project, forcing developers to pay a registration fee to allow the university to place apps on the store. If a web app is hosted internally, however, there will still be costs for hosting and bandwidth.

The problems

The main problem, whether vendor, in-house, native or web app is data. An important aim for us at the University of Stirling was to discover what data would be required from University systems in the long term for mobile apps and whether this data would be difficult to obtain. We discovered how difficult this could be.

The app itself is a shell to consume data from university systems, but obtaining access to this data in a useful, digestible format can cause difficulties. Vendor-supplied products require data feeds in specific formats to plug into their existing app. Accessing this data and creating these feeds in the correct format can take longer than it takes to develop the app. This is due to issues of incompatible systems, systems that sit behind vendor-locked brick walls, lack of APIs or data that is not available in any useful format. For example, one of the most used features of the University of Stirling's app, mStir, is bus times, but these times were not available on any website in a digestible, accessible format such as RSS. Thus, the bus times are manually fed into a custom database and checked for accuracy monthly by a member of staff.

Getting the app used may be more difficult than one expects; the target audience may need encouragement to use it. A major problem with the development of an app at a university is the attempt to boost the app's usage after the initial 'flurry' around the release. Infrastructure can sometimes influence this usage, especially in large university buildings where internet or phone signals can be difficult to obtain or maintain. An example of this decisive barrier to usage is network connectivity. At many universities, users are required to first connect to a VPN (virtual private network) before being able to connect to the internet. The process of a user having to: (1) connect to the Wi-Fi, (2) connect to the VPN, (3) open the app, is often a step too far and users may think that the app is not worth the effort.

Conclusion – the future

For future development, the University of Stirling is taking a step back from the app and focusing on data. The University is attempting to create a web service black box that accesses many of the University systems. The web service will accept authenticated requests for a variety of data and then provide the data in a readable standard for consumption. This rationalized data can then be used not only for any apps, but also for combining a variety of information to feed into other systems for the future. Once this web service has been created, the University aims to create a new in-house app that uses the web service for content population. The app will use frameworks that can be developed using standards such as HTML5, but that also allow the quick, easy creation of native apps to fulfil the expectations of the target audience.

Notes

1 www.mobilestatistics.com/.
2 www.mobilestatistics.com/mobile-statistics/.
3 www.stir.ac.uk/is/student/mobile/email/.
4 http://mwf.ucla.edu/?products/view/1.
5 http://mollyproject.org/.
6 http://kurogo.org/home/.
7 https://github.com/MIT-Mobile/MIT-Mobile-Web.
8 http://mobilewebosp.pbworks.com.
9 www.sencha.com/blog/the-making-of-fastbook-an-html5-love-story/.
10 http://dev.w3.org/html5/webstorage/.

Reference

Davey, T. (2010) 'Information Services Satisfaction Survey', unpublished report.

21

Use of mobile phones for library services: the experience of Hezekiah Oluwasanmi Library, Obafemi Awolowo University, Ile-Ife, Nigeria

Bukky Olufemi Asubiojo

Introduction

Generally, university libraries all over the globe are moving to or grappling with the new trends of deploying mobile devices to provide information services. Twenty-first-century library users want information delivered to the palms of their hands. This prompts libraries to go mobile by using the mobile phone as one of the devices for satisfying users' information needs. However, despite the increasing need for mobile library services, libraries in Africa and Nigeria in particular have yet to use mobile phones for effective access to various library services. Therefore, as an information manager, I set out to discover the readiness of the Library to make use of mobile phones for information service delivery. This required ascertaining those tasks that patrons complete repeatedly within the Library that are amenable to mobile library service and assessing the Library's strengths and weaknesses in terms of readiness for such services. Moreover, it was imperative to take a look at the users' preferences for the delivery of services via mobile phones. The mobile phone is here, and it is here to stay, and university libraries have a responsibility to keep abreast of the dynamically changing information needs of their patrons.

About Hezekiah Oluwasanmi Library

The Obafemi Awolowo University (OAU) is a comprehensive public institution established in 1962 as the University of Ife. The University is situated on a vast expanse of land totalling 11,861 hectares in Ile-Ife, Osun State, in south-western Nigeria. It has a student population of over 30,000 and runs a centralized library system.

The Library is named after the second Vice-Chancellor, Professor Hezekiah Oluwasanmi. It has a collection of about 800,000 print volumes and subscribes to over 5000 print journals. Computerization commenced in 1994 with the donation of an ICL 386 personal computer, a detached CD-ROM drive and a printer. TINLIB library management software was later supplied and installed in 1995. The Library set up a Computerization Team and a contract for the computerization project was awarded in 1995. The first phase was the computerization of the cataloguing section and the Library hooked up to the internet in May 1998. The Library website was developed in 2000. In 2005, the Library introduced an electronic library with 50 computers for the use of staff and students at a subsidized rate. Subscription to electronic journals started in 2002.

In 2008 the Library subscribed to Virtual library management software, the funding for this being a grant from the Carnegie Corporation of New York. Six federal university libraries in Nigeria came together to subscribe to the Virtual Library software. With this facility, the Library has a functioning Online Public Access Catalogue (OPAC). So far, about 80% of the total collection has been converted to electronic format and can be accessed through the OPAC. A Wi-Fi network covering the Library area and a small radius around the Library has been installed since 2009. The Library subscribed to e-books in 2011. Many staff and students are using the electronic resources. It can be said at this point that the Library is seriously considering moving its services to the palms of its patrons' hands.

Mobile phone users and usage

The mobile phone is one of the means of communication that have transformed our world into a global village. The device can make and receive telephone calls over a radio link whilst one is moving around a wide geographic area. It does so by connecting to a cellular network provided by a mobile phone operator, allowing access to the public telephone network. Statistics show that mobile phones had penetrated to about 79.86% of the world population as of 2011 (Saylor, 2012). Similarly, the Compulife report column of the *Guardian* newspaper of 18 July 2012 stated that worldwide access to mobile phones, both pre-paid and post-paid, had grown to over six billion, of which nearly five billon are in developing countries. According to Peter Lyons, GSM's Director of Spectrum Policy for Africa and the Middle East, mobile phone penetration in Africa had reached 649 million by the fourth quarter of 2011. In Nigeria, statistics show that 72.3% of the population (101.27 million of 140 million) are mobile phone users (Nigerian Telecommunication Commission, 2012), of which young people form a large proportion.

Mobile phones in academia – a literature review

Interestingly, university students, researchers and lecturers use mobile phones not just for communicating but for accessing information, teaching, income boosting and even competitions. Adomi (2006) studied mobile phone usage patterns among undergraduate students and his findings reveal that 57.1% of the respondents used mobile phones for academic matters. This suggests that a good number of students are likely to demand mobile library services, although his study does not indicate the content of the respondents' academic requirements. Perhaps his respondents were using their mobile phones to find information on class work, lecture times or the availability of lecturers.

As noted by Doyle Friskney (Layton, 2009) faculty and students want to use mobile technologies to meet the demands of university life. A student expects campus systems to always be available; this includes accessing library resources from a mobile phone (Layton, 2009). Moreover, studies show that researchers are using their mobile devices for scientific research purposes. Katzen's (2009) findings reveal that 50% of researchers use mobile devices for reading, 43% for searching for facts, 30% for searching for introductory information and 30% for keeping track of research data. By implication, it is imperative for university libraries, which are responsible for the information needs of all and sundry in the university, to wake up to these requirements. Hence, university libraries should use mobile devices to provide information services and/or selective dissemination of information to researchers and lecturers so as to actively support and improve the quality and quantity of academic productivity.

In addition, Pin, Chin and Lian (2009) assert that there is a huge potential to engage the younger generation by providing mobile services such as catalogue search, book reservation or the ability to download stories. Similarly, the report by James (2009) on his interaction with Keren Mills about academic libraries and the development of mobile services shows that library patrons are keen to have mobile access to OPACs and other information such as library hours, location and contact details. This reveals the trends in information seekers' use of mobile phones to access information.

Pin, Chin and Lian (2009) observed that their patrons were jotting down catalogue search results, which can prove problematic when pen and paper are not to hand. They therefore incorporated the SMS Me feature into their mobile portal. This facility allows patrons to send bibliographic details to their mobile phones in the form of SMS while searching the catalogue, and then refer to them when searching for the materials on the shelves. As a result, their library portal has attracted 26,000 page views, 4000 unique visitors and 4000 transactions per month. Their library has received 800 commendations about its portals since they were launched. This suggests that university libraries need to make their catalogues and reservation systems compatible with mobile devices so that library patrons can access and use these facilities online at any place and at any time.

University libraries can also deploy mobile devices to provide notification services to their clients. Fatoki (2006) states that university libraries can use mobile phones to send renewal notifications to alert patrons that borrowed books are due back, since most of them return overdue books with excuses that they have forgotten the due date. This aspect of mobile library services will definitely be popular with library users.

Katzen (2009) surveyed researchers' opinions on what content they would consider useful to be delivered via mobile devices. His report reveals that 87% of respondents rated abstract databases with full-text links as useful, while 76.5% rated full-text articles, 64.8% rated book chapters and 71% rated RSS/alerts as useful.

Furthermore, university libraries can deploy mobile devices in the information services section of the library. Fatoki (2006) and Iwhiwhu,

Ruteyan and Eghwubare (2010) suggest that libraries can use mobile phone to answer user queries. This means that librarians at the information desk can respond to reference questions without patrons needing to visit the library. According to Iwhiwhu, Ruteyan and Eghwubare (2010), the SMS or text facilities available on all mobile phones can be used to create current awareness among patrons about upcoming events and new arrivals. This means that the library should keep the mobile phone numbers of all registered library users so as to enable the broadcast of text messages to all the library contacts.

In a nutshell, the use of mobile phones for library services is indispensable in today's world of information products and services. Review studies such as Pin, Chin and Lian (2009) and Katzen (2009) show that libraries in the developed world are already using mobile phones to cater for the information needs of their patrons. Hence, my study aimed to:

1 examine users' preferences for accessing library services through their mobile phones
2 identify library services that users would like to have delivered to their mobile phones
3 investigate the readiness of the library to deploy a service to mobile phones.

Methodology

The descriptive survey method was adopted for the study. Five hundred undergraduate students and ten professional librarians were selected to participate, using simple random sampling. Two questionnaires were designed, for professional librarians and undergraduate students, respectively. Five hundred copies of a questionnaire with two questions consisting of 16 dimensions were administered to undergraduate students, while ten copies of a separate questionnaire with four questions consisting of 21 dimensions were distributed to the professional librarians. Four hundred and fifty surveys were completed by undergraduate students – a completion rate of 90%. These were analysed using frequencies and percentages.

Analysis

Table 21.1 shows that all the professional librarians who participated in the study were aged 31 and above, with an equal gender balance and different grades ranging from Librarian II, to Librarian I, Senior Librarian and Principal Librarian and above. Professional librarians in the study also worked in ICT, cataloguing, circulation, reference and serials.

The majority of the undergraduate students who participated in the study were between the ages of 20 and 25. Table 21.1 also shows that 50.9% of the undergraduate students were female, and 92.9% were single.

Table 21.1 Demographics data of survey respondents

Professional librarians (n = 10)				**Undergraduate students (n = 450)**			
		Number	%			Number	%
Age	31–40	5	50	Age	>20	33	7.3
	41 and above	5	50		20–25	351	78.2
					26–30	34	7.4
					31–35	32	7.1
Sex	Male	5	50	Sex	Male	221	49.1
	Female	5	50		Female	229	50.9
Designation	Librarian II	2	20	Marital status	Single	418	92.9
	Librarian I	3	30		Married	32	7.1
	Senior Librarian	3	30				
	Principal and above	2	20				
Unit	ICT	2	20				
	Cataloguing	2	20				
	Circulation	2	20				
	Reference	2	20				
	Serials	2	20				

The study elicited responses from undergraduate students on their preference for information to be provided to them on their mobile phones. About 90% expressed a desire to be served through their mobiles. They were also asked to indicate those resources and services that they would like to access on their mobile phones. The results are presented in Figure 21.1.

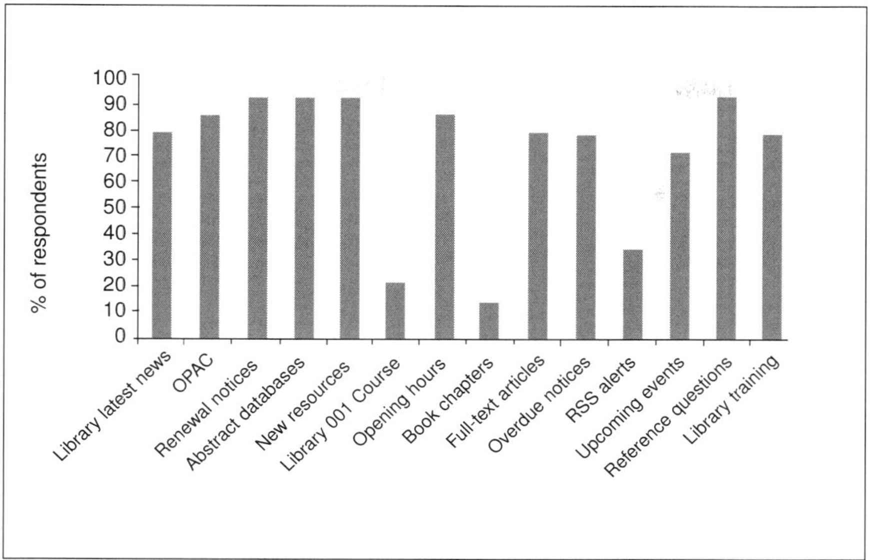

Figure 21.1 Library resources and services that users want to access on mobile phones

Readiness of the Library to use mobile phones to deliver library services

Ten professional librarians were asked to reply either yes or no to a question on the Library's readiness to use mobile phones to deliver resources and services to users. Ninety-nine per cent answered in the affirmative. They were also asked to identify the facilities that the Library already has for the deployment of mobile-phone enabled library services. They all identified the Library website, web-based OPAC, stable power back-up, human resources, electronic information resources and student ownership of smartphones. They were also asked to identify challenges that they felt could militate against the delivery of library services to mobile phones. Figure 21.2 overleaf shows the result.

Discussion of findings

The study revealed that a majority, 90%, of undergraduate students would prefer to access library resources and services on their mobile devices. The fact that 78.9% were between the ages of 20 and 25 shows

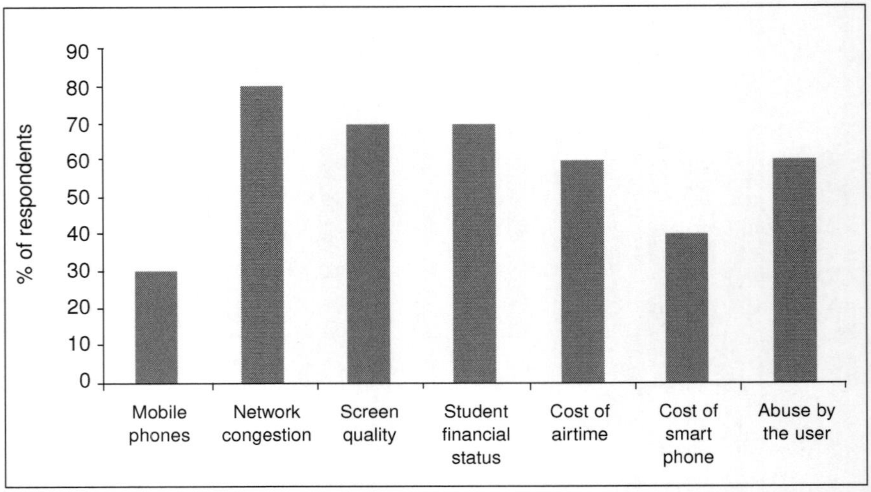

Figure 21.2 Challenges for the use of mobile phones to deliver library services

that the younger generation preferred to use mobile phones to meet the demands of academic life. This result corroborated Pin, Chin and Lian's (2009) assertion that there is a huge potential to engage the younger generation by targeting mobile library services to them.

Moreover, the study showed that 92.7% of the undergraduate students would like their reference questions to be answered through mobile devices. This means that they could be in their halls of residence, lecture rooms or even away from campus and still receive reference service. This could also mean the reference librarians working around the clock, as these young people could need information at any time. The survey also revealed that 92.1% of undergraduates would like to receive renewal notices and to access abstract databases and lists of new resources on their mobile phones. This result indicated that the majority of undergraduates do not want to break library rules by keeping borrowed materials beyond the usual loan period. Also, the number of library defaulters would be likely to fall with the use of mobile phones, since 78.2% indicated that they would like to receive overdue notices, which might prompt them to return overdue materials.

Users were also interested in abstracts and new resources acquired by the Library, which was a good sign that the Library's resources are relevant to their information needs. About 85.0% would also like to access the OPAC on their mobile devices. This would necessitate activation of the mobile feature of the OPAC module of the Virtual library management system. It is of interest that 85.6% of the undergraduates would like to receive information on Library opening hours on their mobile devices. This may be because of changes in opening hours during emergencies, such as when there is a crisis on campus. The survey findings supported the idea that library users want their information needs to be met via their mobile phones.

The study considered all the available facilities that would be needed for the deployment of mobile library services. The findings showed that 90% of the professional librarians surveyed thought that network congestion would be an inhibiting factor, while the availability of mobile phones was the least of the problems. This may be because professional librarians felt that the procurement of mobile phones for Library use should be within the reach of the Library. The financial status of the users, screen quality and cost of airtime were also considered problematic. This is understandable, since many students are not earning but are financially dependent on their parents and guardians.

Conclusion

The study revealed that the users would prefer to access library resources and services through their mobile phones. The majority of the librarians surveyed felt that the Library was ready to use mobile phones for library services in that the facilities existed to support the service. Although some inhibiting factors were identified, these were surmountable.

Recommendations

- The Library should make its web-based OPAC compatible with mobile devices.
- The Library should request mobile phone service providers to increase the capacity of their networks so as to curb network congestion.
- The Library, through the University, should partner with willing mobile service providers to reduce the cost of airtime for all transactions within the university campus.

- Users would need to be trained in the correct way to use mobile services, so as to avoid abuse of the service.

References

Adomi, E. E. (2006) Mobile Phone Usage Patterns for Library and Information Science Students at Delta State University, Abraka, Nigeria, *Electronic Journal of Academic and Special Librarianship*, 7 (1), http://southerlibrarianship.icaap.org/content/ v07no1/adomi_ e01.htm.

Fatoki, O. K. (2006) Prospects of GSM Technology for Academic Library Services, *The Electronic Library*, 23 (3), 266–73.

Guardian (2012) Mobile Phone Access Reaches Three Quarters of World Population (18 July).

Iwhiwhu, B. E., Ruteyan, J. O. and Eghwubare, A. (2010) Mobile Phones for Library Services: prospects for Delta State University Library, Abraka, *Library Philosophy and Practice* (e-journal).

James, C. (2009) Talking with Keren Mills about Academic Libraries and Development of Mobile Services, *Library Connect Newsletter*, 7 (4), www.elsevier.com/libraryconnect.

Katzen, J. (2009) Funding Innovation: is research going mobile? *Library Connect Newsletter*, 7 (4), www.elsevier.com/libraryconnect.

Layton, L. (2009) Talking with Doyle Friskney about How Mobility is Challenging Academic Libraries, *Library Connect Newsletter*, 7 (4), www.elsevier.com/libraryconnect.

Nigerian Telecommunication Commission (2012) *Industry Statistics Subscriber Data*, www.ncc.gov.ng/.

Pin, W. W., Chin, L. Y. and Lian, C. L. (2009) The Library in Your Pocket: making the library truly accessible anytime anywhere, *Library Connect Newsletter*, 7 (4), www.elsevier.com/libraryconnect.

Saylor, M. (2012) *The Mobile Wave: how mobile intelligence will change everything*, Perseus Books/Vanguard Press.

22

If you build it will they come? The importance of collaboration and marketing in developing and promoting mobile services

Binky Lush and Emily Rimland

Introduction

No library mobile site needs be a site unto itself. Rather, collaborating with campus partners and constituents on the development, and taking steps to advertise new mobile services, behoves everyone – especially users – who will benefit from a co-located suite of academic services to meet their needs.

The Penn State University Libraries serve almost 100,000 graduate and undergraduate students, as well as Penn State Alumni, faculty and staff and the general public. In 2009, we saw a growing trend among this diverse group of library users, namely, the increased use of mobile devices to access information.

Gathering needs

With the understanding that mobile usage was exploding, but without a clear roadmap for delivering the Libraries' services to mobile users, we attended the second M-Libraries conference, which helped us to formulate our mobile strategy. Our first step was to survey a cross-section of our students, asking them what library services they would be most likely to access via a mobile device. Not surprisingly, the top three responses included checking hours, searching the Libraries' catalogue and looking for an available computer (Table 22.1 on the next page).

Table 22.1 2009 Mobile library services survey, 2009

Service	Respondents (%)
None	32.7
Check Library hours	32.5
Access the catalogue	25.7
Check computer availability	24.3
Access Library account	20.5
Look up Library contact information	18.7
Ask a research question/get help	18.6
Access a Library database	17.0
Get Library news and events information	7.1
Other	0.5

Figure 22.1 First mobile site, launched in 2010

Based on our findings, we planned to provide a site that was limited in content, concentrating our development on tasks that would be performed when our users were 'on the go'. In the spring of 2010, we launched a very basic text-based library site, as shown in Figure 22.1.

Partnering to develop a mobile site

Although this was a valiant first step, there was no denying that our site was rudimentary in comparison to the device-specific applications being released at the time. In addition to upgrading the functionality and look and feel of our first mobile site to provide a more 'app-like' experience, we hoped to integrate the Libraries' mobile site with other University mobile services. In early 2011, the Emerging Technologies (ET) Department at Penn State was tasked with developing a University-wide mobile presence built on the MIT open source Kurogo framework, and needed a partner with whom to collaborate on developing a prototype for data sharing. The timing was perfect for the University Libraries. We understood what Libraries users needed – mobile access to the Libraries' resources, within the context of the University's mobile site. We provided access to our search, computer availability data and library locations and hours information and worked with ET to develop a process for sharing data across information silos, helping to determine how this data could be integrated and mashed-up with other University data sources to offer essential information for our mobile users.

In the fall of 2011, after several months of collaboration with ET, the Penn State University Libraries moved from its own stand-alone mobile site to a site that was fully integrated with the University's mobile presence (see Figure 22.2). This integration afforded the Libraries a new mobile look and feel, better web-scale search capability for Libraries holdings (using Summon) and a number of other advantages, including geolocation services, computer availability within the Libraries, a streamlined hours display, mobile-optimized databases, mobile apps for databases and access to a user's library account and full library website. Future enhancements planned for the Libraries site include text reference and the full A–Z databases list optimized for mobile.

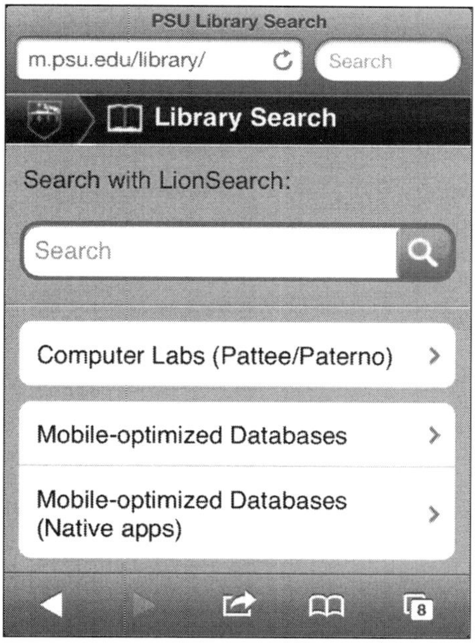

Figure 22.2 Libraries' mobile site integrated with full University mobile site, launched in 2011

The partnership between the Libraries and ET has allowed the Libraries to foster an ongoing relationship with another forward-thinking IT unit at Penn State and to collaborate on sharing data across all campuses and departments for a variety of applications, breaking down the many data silos that exist in a large University. Most importantly, this collaboration has been the start of a solid and fruitful partnership between the Libraries and University IT!

Mobile site launch and marketing

As ET continued its work towards University-wide collaboration and data sharing, m.psu.edu was quietly launched as the University's beta mobile site. The Libraries portion of the site, however, was fully functional and became its production mobile presence. It was time to get the word out to the diverse user base!

Mobile library services, like traditional library services and spaces, should not be reliant upon the axiom 'if we build it, they will come'. Although mobile library sites and services are becoming popular, essential and expected by users, it is important to market these services and reach users at different points of need. While social media and the web are natural avenues to advertise your mobile presence, traditional or print-based materials remain effective tools for spreading the word about your services. Penn State University Libraries executed both traditional and digital marketing plans with the re-launch of its mobile library site in 2011. This strategy encompassed a variety of outlets, which will be reviewed here.

Since the mobile library site was integrated within the University's overall site (m.psu.edu), we expected some large-scale marketing to come from the University's central marketing unit as a 'top down' promotional campaign. However, a 'soft' launch was more befitting the University's new beta site. There was also speculation that the Penn State community would naturally 'stumble upon' the new mobile site, since it adhered to the standard mobile practice of using m.websitename as the URL. These assumptions, as they so often are, were wrong. To date, the University as a whole has not advertised the mobile site broadly to Penn State students, faculty and staff. There are surely some early adopters who found the site by guessing at the URL; however, most people were unaware of m.psu.edu. As these limitations became apparent to Libraries early on, we were able to respond with our own marketing campaign to spread the word about the library section of the mobile site.

Naturally, the first step was to market the mobile library site in the digital arena. This included outlets like social media (Facebook and Twitter), via the Libraries website and digital signage in the library itself. The first step in this digital promotion was via Libraries' internal public relations (PR) and marketing department, which created original artwork that was used in a variety of ways (Figure 22.3). The social media posts were simple, short blurbs about the roll-out of the service, what you could do on the site and the URL. The Penn State Libraries home page (libraries.psu.edu) gets over 6000 hits per day and therefore provides a prominent way to send a message to a target audience. The website has a 'carousel' of rotating images which prominently features programmes, events and announcements. We utilized the carousel to advertise the mobile site using the graphics supplied by the PR department, which linked to the mobile site. We took advantage of the flat-screen displays interspersed throughout the first floor of the Libraries' flagship building, Pattee and Paterno Library. These displays used the artwork developed in-house to advertise the mobile site to visitors of the physical library.

Since the mobile library site was so new, and also separate from Libraries' main website, we felt that we had to advertise the site via more traditional methods so as to cast the widest net possible. The PR department also created a press release, posters and large 'sandwich board' placards for display inside and outside the library, an ad in the student newspaper and an article in the Libraries' internal newsletter. A QR code which was a shortcut to the mobile site was also incorporated into many of these items. Additionally, librarians began telling their faculty and students about the site, since there were many helpful services available and this word-of-mouth source was not to be discounted. Once these marketing devices were employed, use of the mobile site did increase, indicating the effectiveness of our promotional efforts.

Figure 22.3 Artwork developed by Libraries' Public Relations department to advertise the new mobile site

Conclusion

Although it may seem counter-intuitive to market a mobile service via analogue means such as posters, ads and press releases, our experience showed that traditional print marketing is valuable and vital to launching a new service such as a mobile website.

The mobile landscape is changing rapidly, and the Libraries must keep up in order to meet users at their point of need. The Libraries' collaborative development of the mobile site and our marketing efforts using both digital and analogue mechanisms were a great step into the mobile realm and allowed us to provide and promote our Libraries' services to a wider and increasingly mobile audience.

23

Delivering news on mobile: the European Parliament's m.Library website

Caroline Corneau

Introduction

The Library of the European Parliament (EP) provides information services to support Members' legislative activities. Its intranet site is used by 754 Members and their offices (1500+ assistants and 1000+ political groups staff), and 5000+ Secretariat staff, as the main access point to Library information and services. These include the Library catalogue, electronic documents, the Library's own branded products such as briefings and research guides, as well as information services like the enquiry service 'Ask the Library' and 'News', which gives access to news agency feeds.

The m.Library project and development

In response to requests from Members who wanted to view Library information on their smartphones and tablets – and, above all, to read breaking news – a pilot project was launched in March 2011 with the aim of delivering Library services on mobile devices. As a first step, a mobile website called m.Library providing real-time access to news bulletins was developed in-house, based on the existing 'News' service from the desktop site. A prototype was produced by June 2011 and was tested by a selection of 30 Library clients. The Library's mobile website was then launched in October 2011.

The m.Library website was built in a development time of 18 weeks by one developer (10 weeks for the prototype and 8 weeks for improvements before the launch). It is a mobile-optimized website compatible with HTML5, JSP and CSS3. It uses Java EE, Spring MVC, Hibernate and SiteMesh frameworks. Only standard technologies validated by the EP's IT department were used, as required by the site's location within the EP Intranet and its secured access via the EP's extranet portal.

The m.Library website: content and features

The m.Library website offers newsfeeds from nine national and three European news agencies, published in real time. National agencies' feeds are updated every ten minutes, while the European agencies publish daily bulletins. Over one million individual articles are published in a year, with more than 5000 daily on weekdays.

Site structure

Compared with its desktop counterpart, the layout and functionality of m.Library are simplified and tailored for use on mobile devices. The site's home page (Figure 23.1) is very simple to allow for the addition of future services. It consists mainly of icons presenting the available sections, with the possibility to add a short communication message for marketing purposes. The user's first and last names are also displayed: this helps to remind the user that the site's content is restricted to EP staff, but also hints that, just as for the desktop site, some personalization features are available. From the home page the user can

Figure 23.1 The m.Library home page

choose between the 'News' section and an 'About' page which presents general information about the Library, such as contact information, location and opening hours.

News section features

The number of features available on m.Library, as well as specifications for navigation and saving profile preferences, are adapted to mobile use. The following is an overview of the main elements.

Browse

When 'Read the news' is selected, the section displays the 25 latest newsfeeds based on the user's selection of agencies or languages (see 'Preferences' below). Feeds are always displayed in a last-in-first-out order. At this 'Newswires' list level, the information displayed for each feed is limited to the title and first lines of the article, the name of the agency, and the date and time of publication (Figure 23.2). The user can scroll the list of articles, tap on any item to read the complete text, and from there browse to the next or previous articles, or go back to the list. The user can also choose to view older articles by tapping the 'Load more' button at the bottom of the list, which loads 25 more items each time. Compared with the desktop site, both the number of items and the amount of information are reduced, and there is no calendar for browsing by date.

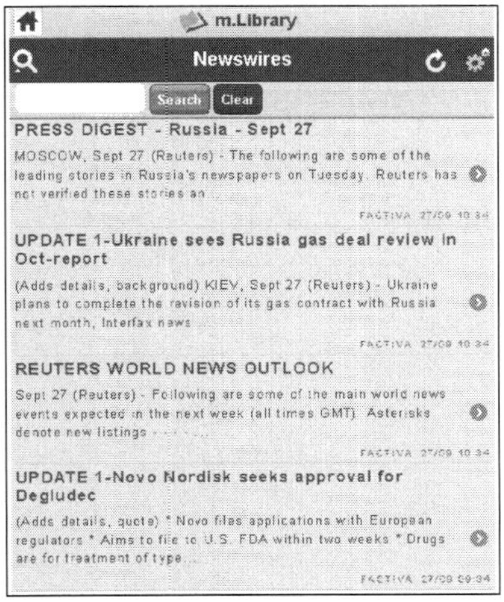

Figure 23.2 m.Library News section

Search

A simple, collapsible search box is provided at the top of the Newswires page (see Figure 23.2). When a search is submitted, the articles are immediately filtered to match the search criteria, but the order remains reverse-chronological. The search filter can be deactivated by a 'Clear' button. Comparing with the desktop site again, no advanced search form or faceting of results are provided, nor are there any options for changing the order of the results.

Share

Users can send newsfeed links (on both mobile and desktop sites) via their personal e-mail, from where they can select their contacts. This is similar to what is available on the desktop site. For copyright reasons the full text of articles may not be broadcast to non-EP staff users.

Preferences

Newsfeeds can be filtered either by agencies or by languages. By default, the values are based on the user's selections on their previous visit to either the desktop or mobile website. On a first visit, no selection is activated. Values are synchronized between the mobile and desktop sites.

Refresh

The application does not refresh automatically during a session, as the user's internet connection might not be continuous. The user thus needs to click on Refresh in order to download the latest updates. However, each time the user reconnects to the application, the latest newsfeeds will be displayed.

Best practices in mobile design and development

It is beyond the scope of this chapter to define guidelines for mobile development, and many sources are already available on that topic. Relevant best practices in mobile design and development were identified and applied to create a mobile-optimized product. These also allowed the project team to devise an analysis grid for evaluating mobile news services. The following are some key elements that helped to shape the m.Library website (Figure 23.3):

Follow standards from similar websites, as users will be influenced by current practices when gauging the quality of a new website.
Limit the number of features to those most relevant to mobile use.
Simplify the browsing scenario as much as possible.
Harmonize the layout and working mechanisms of all pages.
Plan for the future: design the home page as set of icons displaying the list of available services.
Follow a 2-level navigation approach: a first level for general navigation and a second level for the actions related to each section.
Synchronize data, functions and settings with the desktop website.

Figure 23.3 Different steps in the design of the m.Library website

Technical constraints

Secure access via the EP extranet portal (EPNET) played a significant part in the decision to develop a website rather than an app, and had important consequences for usability. Due to contractual terms, news is available only to EP staff, who therefore have to log in to the intranet to access the feeds. The EPNET portal is the only authentication method provided by the EP's IT infrastructure for access from the internet, so it is essential when using a mobile device. In granting access to the Intranet, the portal rewrites the HTML code of the internal applications, so only the most basic technologies could be used, excluding the use of HTML5 mobile application frameworks or a native mobile application. The authentication process itself had a negative impact on the site's usability, since it requires a manual login each time a user connects (no auto-save is allowed) and allows a limited session duration of only 30 minutes.

Client feedback

Comments on the initiative of launching mobile Library services and on starting with the 'News' service were mostly positive. The design of the prototype was also commended for its logic and ease of use. Criticisms and suggestions for improvements focus on four areas:

- The login procedure should be omitted or simplified.
- More news databases or newspaper content should be made available.
- Some functions could be improved or added, such as an alert system or thematic filters.

- A real native mobile application would be more user-friendly than a web version.

New developments in the Library's mobile strategy

Client feedback and lessons learned in terms of the feasibility of mobile services supported a recommendation to continue development as part of the Library's mobile strategy. A mobile version of the 'Ask the Library' service was developed in 2012, to be launched in early 2013. A search facility for the Catalogue and all Library products and a mobile-friendly version of the existing Alert system are also being considered.

However, within the framework of its New Concept 2011–13 programme to increase the value of its services to Members, the Library has now launched a project to redesign its website with the aim of providing an improved and simpler online experience to its clients, including online information services across platforms (internet as well as existing Intranet) and devices (desktop, tablet, smartphone). This new website should be launched by mid-2014 and will make use of responsive web design.

Within the EP, close contacts with the IT and Communication departments working on different mobile projects were initiated in the first phase of the m.Library project, and these will be maintained in order to share best practices and knowledge on both in-house and outside mobile developments and co-ordinate efforts for the benefit of our mutual clients. In particular, the Library is involved as a key user in projects to improve access and login procedures to the Parliament's internal resources and to the Wi-Fi service provided within the Parliament.

Acknowledgement

Thanks to Messrs Alfredo De Feo, Director of the Library, and Iain Watt, Head of the Strategy Unit, for their support and guidance in the m.Library project, John Cardinal who single-handedly developed the website, Mitja Brus as the finest mobile stakeholder/key user, and Ron Davies for his encouragement and the best team partnership.

Note

This text is written in a personal capacity and does not necessarily represent the views of the European Parliament.

24

Searching the Library catalogue through Twitter

Bianca Kramer

Introduction

In the autumn of 2011, the *Guardian* newspaper started an experiment called the Guardian Tagbot, an automated service on Twitter. It enabled people to send a tweet to @GuardianTagbot (please note: this Twitter account is now defunct) with a query and to receive a tweet in response with a link to a list of *Guardian* articles on whatever subject is specified.

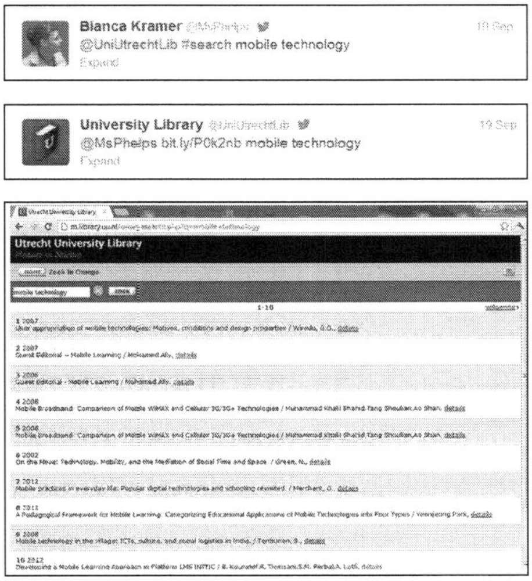

Figure 24.1 Example of Utrecht University Library's Twitterbot (a) User query, (b) Twitterbot response to user, (c) Query results from online Library catalogue

While working at Utrecht University Library, I was inspired by this example of Twitter use and attempted to construct our own Twitterbot, which would respond to a user's queries by sending a link to results from the Library's online catalogue (Figure 24.1).

Not being a programmer, but having an interest in the possibilities of web-based technologies, I developed this Twitterbot using freely available web-based tools. My journey started with a Google search for 'Twitterbot without coding', which quickly pointed me to a blogpost on Digerati Marketing, titled 'How to make a Twitter bot with no coding'.[1] I took my cues from there and set to work, adapting tools and techniques as needed.

From RSS to RSS

The Twitterbot workflow is built around RSS feeds. First, an RSS feed is generated from tweets to the Library's Twitter account containing a search query, identified by the predefined use of the hashtag #search. Subsequently, each item in the feed is modified and a second RSS feed is generated containing the modified items, which are then sent as individual tweets by Twitter. To accomplish this, I used two freely available web-based tools: Yahoo! Pipes (http://pipes.yahoo.com) and IFTTT (If This Then That, http://ifttt.com).

Yahoo! Pipes

Yahoo! Pipes is a powerful tool for modifying RSS feeds. To generate an input feed (in this case, an Atom feed, which is similar to an RSS feed), Twitter's search functionality was used to construct a URL[2] which pulls up all tweets sent to the Library's Twitter account (@UniUtrechtLib) containing the hashtag #search.

Yahoo! Pipes allows a modular approach to modifying items in the input feed, wherein different 'building blocks' are used to accomplish different tasks. In our case, we first eliminated any retweets (RT) from the feed, to prevent our Twitterbot from giving duplicate answers. Then we stripped the hashtag #search and our own Twitter account's name from the body of each tweet, leaving only the search terms used in the query. These were then incorporated into a custom URL representing the search results from the online Library catalogue.

To illustrate how this 'URL-hacking' works, take the example query 'mobile technology'. When this term was searched for in the mobile version of our online library catalogue (as it was in use at the time), the resulting URL was:

http://m.library.uu.nl/omegasearch.php?q=mobile+technology

Taking the stem of this URL:

http://m.library.uu.nl/omegasearch.php?q=

any search terms could be added to generate a URL for the results of that query. Boolean operators, brackets and quotation marks in the search query can also be incorporated into the URL.

Each modified tweet in the RSS feed now contained the original search query and the custom-made URL for the query result. In a final step, the Twitter account name of the sender of the tweet was added to the body of the tweet, to make it into a reply to that user.

Yahoo! Pipes made these modified tweets into an RSS feed which could then be used to send the actual tweets via Twitter. For this, the second tool, IFTTT, was used.

The complete Pipe is public and can be viewed via its own URL.[3]

IFTTT

If This Then That (IFTTT) is a beautifully simple tool that enables users to specify that if any given thing happens on the internet ('If This'), some other action is initiated ('Then That'). There are many different input and output channels, and possibilities (called 'recipes' at IFTTT) range from 'When you post a photo on Instagram, share it on Facebook' to 'Post events from Google Calendar to Twitter'.

In our case, it was configured in such a way that every time a new feed item occurred in the RSS feed from Yahoo! Pipes, the item was sent as a tweet from the Library's Twitter account.

The complete recipe is public and can be viewed via its own URL.[4]

Limitations

Using the techniques and tools described above, we made a functional Twitterbot. However, due to the characteristics of both Yahoo! Pipes and IFTTT, there was a major drawback in the Twitterbot's functionality, namely its response time. This is dependent on the update frequency of both Yahoo! Pipes and IFTTT, and proves to be between 0 and 15 minutes, with an average of about 8–10 minutes. While this might not be problem for an answering service using a different platform (such as e-mail), it is a major issue for a service using Twitter, where near instantaneous responses are the norm.

Given this limitation in functionality, the Twitterbot was launched as an experimental service and we kept marketing efforts to a minimum. The Twitterbot attracted attention from librarians from the Netherlands and abroad; however, its use by Library patrons was limited. Whether this was due to unfamiliarity, dissatisfaction or lack of interest in such a service is currently unknown.

The Twitterbot has been 'live' since November 2011. However, its continuing deployment is hampered by recent changes in the Twitter API. As from 5 March 2013, Twitter support for Atom and RSS ceased, and only JSON is supported. In addition, use of the Search API now requires authentication. Continuing use of the current Twitterbot set-up will thus require changes in the way Yahoo! Pipes is used.

Possibilities

Despite the above limitations, we feel that the Twitterbot offers interesting possibilities, mainly as proof-of-concept for employing relatively easy-to-use web-based tools to develop new services.

If one accepts the drawback of the lag in response time, the Twitterbot can easily be adapted to search other databases, both other libraries' catalogues and more general databases that allow URL-modification to generate search results. Examples are PubMed, WorldCat and Google Scholar. Several web-scale discovery systems like Primo and Summon also fit this kind of service.

On a broader scale, such services do not need to be limited to Twitter and/or searching. IFTTT supports a wide range of input and output channels, providing the opportunity to develop many kinds of services. Some may be useful to facilitate behind-the-scenes activities, such as sharing promotional library photos or event announcements across different media, while others may facilitate direct interaction with patrons, like our Twitterbot.

Conclusion

Using web-based tools, such services can be developed with relative ease, either fully functional or as proof-of-concept prototypes. This would allow librarians to quickly test new ideas without having to rely on the support of software developers within the organization, who are often not able to accommodate ad hoc requests.

If an experimental service proved to be successful, or potentially interesting enough to merit further development, the software development team could then be asked to script all or part of the workflow and to incorporate the service into the organization's website. Scripting and hosting the service in-house instead of relying on web-based tools is a way to overcome the drawbacks of these tools, such as the lag in response time described for the Library's Twitterbot. It would also allow for further customization of services.

An interesting initiative in this regard is Pipe2Py, a code generator that automatically translates the functionality of a Yahoo! Pipe into Python code. For those who know how to handle a Python script, it is available on GitHub.[5] In addition, it is also possible to create and run the Python script online using pipes-engine, an application with Google Apps Engine that uses Pipe2Py.[6]

Applications such as these, although they go beyond the level of 'no coding', might also be of use in further developing new services for library patrons.

Notes

1 www.digeratimarketing.co.uk/2009/09/23/how-to-make-a-twitter-bot-with-no-coding.

2 http://search.twitter.com/search.atom?q=%23search+to%3AUniUtrechtLib.

3 http://pipes.Yahoo!.com/ubtestbot/uniutrechtlib_direct.

4 https://ifttt.com/recipes/77022.

5 https://github.com/ggaughan/pipe2py.

6 http://pipes-engine.appspot.com.

25

Mobile devices in medical schools: the WCMC-Q experience

Sa'ad Laws

Introduction

Since its release in 2010, the iPad has revolutionized the way in which clinicians are able to provide medical care. Clinicians can now access information that is needed to improve patient care and recovery time. This development has raised the expectations of library users that we can assist them in identifying, evaluating and accessing mobile resources, but it has also allowed us to engage with our users in new and progressively dynamic ways.

The following case study looks at how this shift is occurring at Weill Cornell Medical College in Qatar (WCMC-Q) and, more specifically, within the college's Distributed eLibrary (DeLib). WCMC-Q, like many American medical schools, is providing its medical students with iPads. While getting these devices into the students' hands is a challenge, finding ways for the students to use them effectively is possibly an even more daunting and demanding matter.

Why use iPads at WCMC-Q?

The main origin of the iPod and iPad project was to mirror our sister programme at Weill Cornell Medical College in New York; however, due to many other considerations, it made sense to institute this programme in Doha. WCMC-Q, like many other medical programmes, is always looking for more efficient and innovative ways to engage with and instruct its students. Foremost is the desire to engage students within a progressive framework.

WCMC-Q also had some logistical interests, such as the problem of ordering books from the US. To resolve this problem, WCMC-Q has worked hard to identify and purchase electronic materials, such as e-books. Furthermore, the use of electronic materials has made the dissemination of course materials to students much easier, as they can now gain access to the materials via e-mail or through a course management system.

One of the greatest benefits of the programme has been its impact on the environment, by eliminating the need to print and distribute multiple copies of course materials or to purchase printed textbooks. Many other schools have found that the reduction in printing costs alone can often compensate for the cost of the iPad devices.

Mobile devices at WCMC-Q and DeLib's involvement

Beginning in 2010, WCMC-Q distributed iPods to third-year medical students. DeLib's initial role was as a consultant, recommending apps for medical students. At that time, the Library recommended several apps on the basis of their general applicability, such as the Skyscape App suite. At that time, there was no emphasis on clinical application, but rather on general medical reference use.

After the College made the switch from iPods to iPads in 2011, a PDF annotation app, as well as other productivity apps, was included in the package of apps that students received. It was at this point that DeLib began to expand its role within the College's overall iPad plan. In addition to advising the Medical Education department on the selection of specific apps, the Library was asked to conduct introductory workshops for first-year medical students. These workshops instructed students in the use of core medical apps as well as several productivity apps.

In addition to initial instructional efforts, DeLib librarians were involved in administrative aspects of the College's iPad programme. Librarians worked with individual faculty members to develop plans for incorporating mobile medical resources into their courses. Librarians were also asked to make presentations on various mobile resources topics at the College's clinical curriculum meetings. All of these efforts have not only increased DeLib's credibility and visibility within the

College, but have also given it greater authority in the development of the programme.

Based on these early experiences with students and faculty, DeLib decided to take a more proactive course of action in order to integrate information literacy and technical skills building into its educational and reference initiatives. DeLib began by developing a Mobile Resources web page that identified apps and classified them by genre, such as calculators, medical reference, etc. The page offers various 'tips and tricks' for users to help them to customize their devices and make them more functional. Since its inception, the web page has moved to a LibGuides format and is currently the most-viewed page within DeLib's LibGuides content.

To generate dialogue among students, DeLib began the 'Student App of the Month' programme. Students are asked to submit apps that they find useful in their academic pursuits. These apps are not restricted to medical apps, but students should make some connection between their nominated app and their academic studies. Winning entries are highlighted on the Mobile Resources web page, including a short quote illustrating how the app helped the student. To encourage students to submit entries, the Library offers an incentive, typically an iTunes gift card.

DeLib conducts several workshops aimed at introducing iPads and apps to various segments of the College. Examples of workshops include 'Apps for Point of Care', 'Medical Apps', '40 apps in 40 minutes' and others. The workshops are intended to give users some familiarity with the variety of medical apps available, as well as some examples of how they can be used in clinical settings.

Finally, and perhaps most importantly, DeLib has worked diligently to build solid relationships with faculty members. Liaison librarians have worked with key medical faculty to identify different ways that students can use mobile devices to engage with course materials. The efforts that librarians and faculty have made to enable students to use their iPads during hospital rounds are a good example. Librarians identify apps and instruct students in their point-of-care use before attending rounds. While attending rounds with students, librarians help the students to identify appropriate apps and information resources for individual clinical cases.

Challenges and successes

The Library's work thus far has met with many successes and challenges. Most satisfying is the progressive level of dialogue that has been generated within the college. Some of these conversations are about instructional integration, others about clinical best practice and yet others about social and generational differences. However, all of these discussions centre on how to move forward in accepting and promoting mobile device and resource usage in daily practice.

DeLib has attempted to facilitate greater information literacy through integrating iPads with instruction in Evidence Based Medicine (EBM). DeLib participates annually in a series of introductory EBM workshops for first-year medical students. These workshops introduce students to the basic concepts of EBM, such as developing background questions. When dealing with background questions, students are encouraged to identify and use apps to answer their questions. This process demonstrates the utility of mobile devices for quickly accessing precise information in clinical scenarios.

However, despite some preliminary successes, WCMC-Q's faculty have not established any co-ordinated methods for using mobile devices and resources during classroom instruction. Some suggestions have involved the use of social media to facilitate in-class discussion, but many faculty worry that the devices can become more of a distraction for their students than an aid.

Other challenges that have arisen are the logistics of using a US-based product and applications outside of the US. Many faculty and students prefer to use American Apple App Store accounts, and simple issues such as attaching a credit card to a user account can present problems. These and other problems, such as poor Wi-Fi in some satellite clinical locations, require the College and DeLib to be innovative in finding solutions.

The future

For many users, one of the areas of greatest need is finding reliable sources of reviews and recommendations of apps for medical use. Navigating through the maze of apps can be daunting for any user, especially for students. Although many medical libraries recommend apps, they do not offer critical reviews. While there are some websites that provide this service, they are few in number, and more viewpoints would serve users better.

Finally, one area that is essential for any institution with mobile device users is to find effective and meaningful ways of polling users on their needs and behaviours. In order to learn what our users (medical students, in DeLib's case) are doing, it is important to gather not only quantitative information, such as through surveys, but also qualitative information, such as through focus groups and observation. One of the best methods that DeLib has found is to observe students on their rounds and actually see what they are doing with their iPads and what kinds of resources they are using to answer clinical questions.

Conclusion

Mobile devices and resources present a myriad of opportunities for libraries and their users. In some cases, libraries need to find new ways to do old things, such as accessing information. In other cases, libraries are charged with developing new pathways altogether, such as finding effective ways to use mobile devices at the point of care. However, all of these challenges present an opportunity to move forward and meet users where they are living and working. In the case of medical students, this is especially important, given the hectic nature of their studies and their limited access to desktop computing. With mobile resources, librarians have an opportunity to extend their reach and to renew or increase trusting relationships with their users.

26

Kindles in the library: National University of Ireland Maynooth Kindle pilot

Louise Saults

Introduction

With the new era of mobile technology and changes in how users access library resources, the Library at National University of Ireland Maynooth wanted to investigate ways of enabling users to borrow multiple core text materials on a single device. We looked into various options and chose the Amazon Kindle because it was less expensive than the iPad, the cost of the e-books was very reasonable, the device was easy to use and its size made it perfect for students to carry around with them.

Kindle pilot

The Library purchased 10 Kindles in May 2011 and uploaded them with free classic books, with a view to running a pilot project. Library staff had an opportunity to borrow the Kindles over the summer months, which proved very useful, especially for the Reader Services staff, as they would be dealing with any queries from the students. The Library then invited the departments of History, Business and Law to take part in the pilot, as they had expressed an interest in developing new ways of accessing Library resources. These departments were asked to browse the Amazon website and choose titles that would be relevant to their courses. Each department submitted a list of titles which the Library purchased via its Amazon account, the cost of the e-books being charged to the departmental book fund. The e-books were then uploaded to the 10 Kindles. To upload the e-books to the Kindles it was necessary to log on to the Wi-Fi or use the USB cable provided with each device.

Some publishers impose limits on the number of devices to which a copy of an e-book can be uploaded. This restriction means that additional e-book copies need to be purchased, depending on the number of Kindles to be loaded. However, this was still very good value for money.

The next stage in the process was to create a record for each Kindle device in the Library Management System. Each Kindle was given a barcode that was linked to this record. The purpose of the barcode was to allow students to borrow the Kindle. We also created a catalogue record for each e-book on the Kindle, with a note stating where the students could find the Kindles in the Library. Students could borrow a Kindle for one week and the device could not be renewed, which is the Library's standard practice for equivalent printed 'core' texts. Additionally, there was a fine for the late return of a device and a maximum fine of €100 for damage to a Kindle. Each Kindle was kept in a plastic folder, together with its USB cable, instructions on use and a date label, and the devices were held at the issue desk. When a Kindle was returned, the contents of the plastic folder were checked.

The pilot required the involvement of other departments within the Library. The Systems Department created an Excel spreadsheet to document information on each Kindle, such as serial number, barcode, Kindle name. As we had to return one or two Kindles, due to malfunctions, it was important to have a record of this information. The Systems Department also set up a record for the devices on the Library catalogue. The Cataloguing Department was consulted regarding the best record layout. The Acquisitions Department had a major part to play, as it was involved in the purchase of the Kindles and e-books. It monitored the book budget and Visa statements and the constant maintenance and upkeep of the Kindles. And finally, the Reader Services Department was involved on a continual basis, lending out the Kindles and answering any queries from students.

The pilot began in September 2011 and finished in January 2012. It was marketed on the Library website, via Facebook and Twitter and in the Library. There was also an online tutorial about the Kindle on the

Library website. During the course of the pilot users were surveyed and feedback was extremely positive. There was also a Kindle as a prize at the end of the survey.

Survey

The questions asked in the survey and the results are shown in Figure 26.1.

The following are a sample of comments from the survey:

> Very easily transportable – cuts down on the need to carry around heavy books.
> Loan period sufficient for demand.
> Great idea. Much better then reading on a computer.
> I like that the screen is not shining brightly. It's close to a real book!

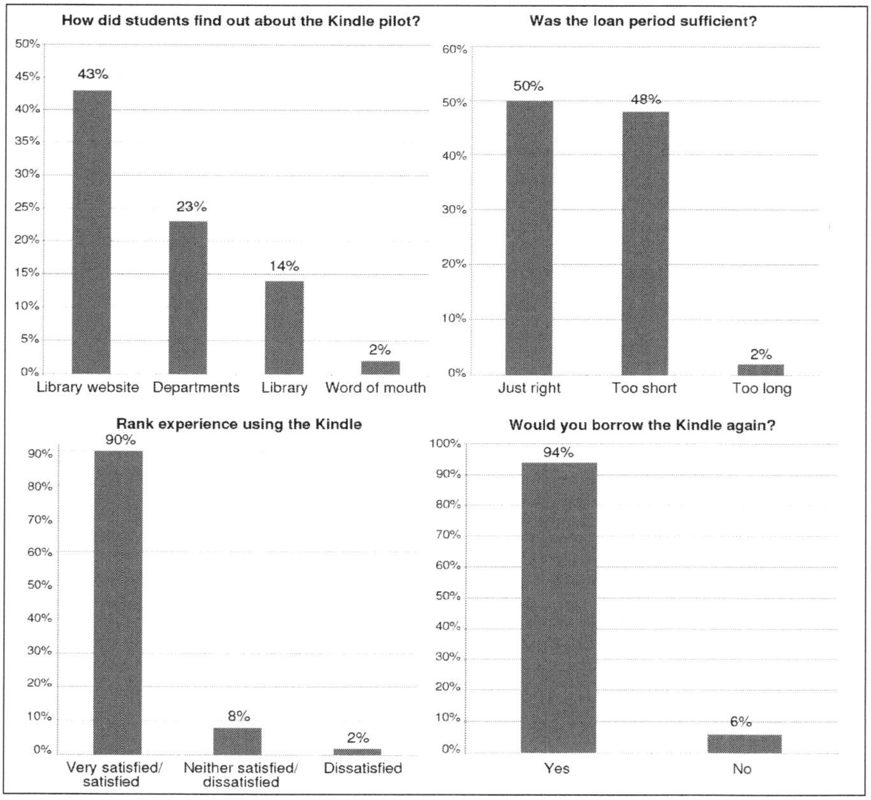

Figure 26.1 Kindle survey and results

> Great to read, very pleased, would use again or buy own – dept textbooks great idea.

As the pie chart in Figure 26.2 shows, the majority of the Kindle loans were to undergraduates and postgraduates. The Kindles were the most popular with first-year Arts students. This was closely followed by second-year Arts and then MLitt research students (see Figure 26.3).

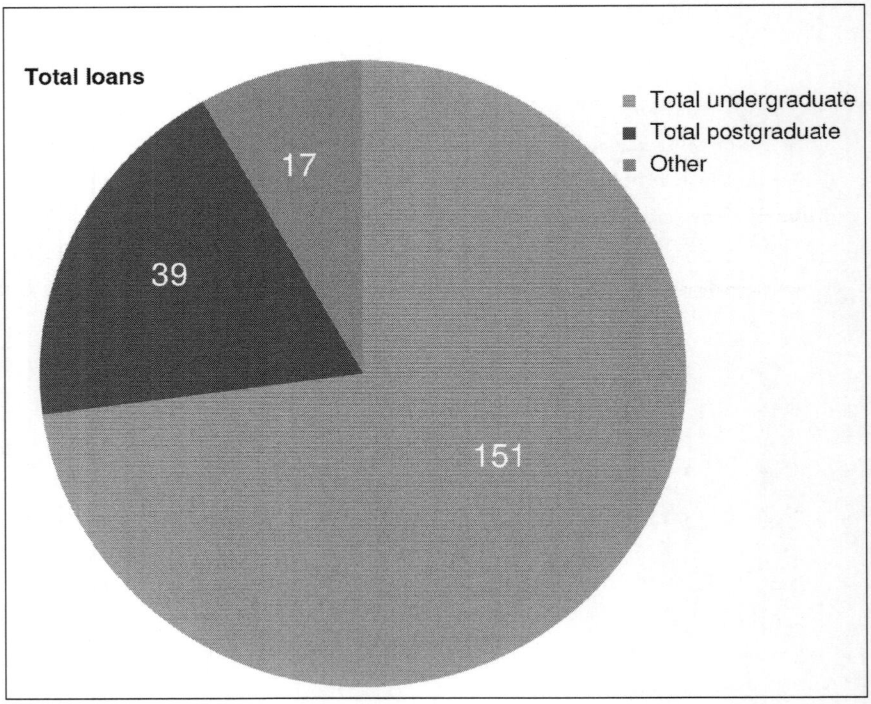

Figure 26.2 Statistics on Kindle loans

Analysis

There was a high percentage of repeated use of the Kindles by students. Out of the total of 174 borrowers, 33 borrowed a Kindle more than once. There was one academic borrower.

Conclusion

As the pilot was such a success, other departments have since become

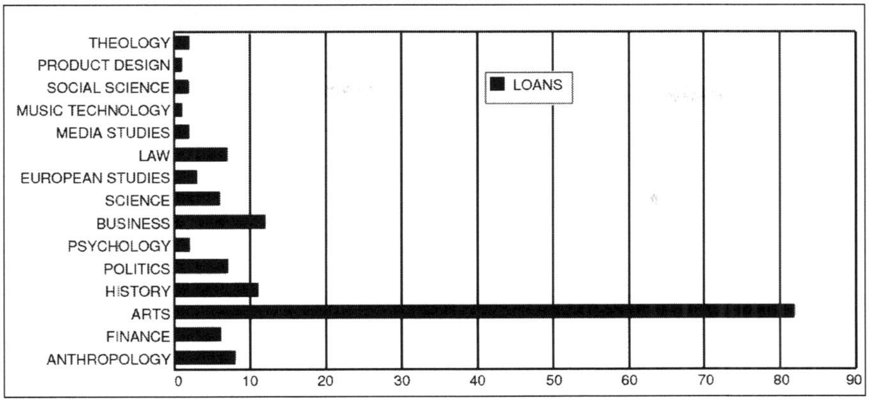

Figure 26.3 Undergraduate Kindle usage, by subject

involved: English, Business, Economics, German and Geography. A total 150 e-book titles were added to the Kindles in October 2012. It is now standard practice for the Library to acquire e-book titles for the Kindles.

Conclusion

The subtitle of this volume is 'From margin to mainstream'. The chapters included here are indeed testament to the extent to which M-Libraries has moved since the publication of the first volume in 2008. While the use of mobile technologies by libraries is not exactly commonplace, an ever-increasing number of libraries have mobile websites, and the range of innovative projects described here is impressive. It is an area where one size definitely does not fit all and libraries are adopting and adapting the technologies to fit their specialist needs and priorities. For some this will be the use of QR codes to help users physically locate items in a vast area of shelving. For others, the use of SMS messaging will address the need to find a quick and effective way of communicating with a geographically dispersed user population, or the use of iPads will free reference librarians from their enquiry desks. In summary, this volume serves to demonstrate the ability of libraries and librarians to identify creative ways of harnessing technology to address problems and to meet the increasingly sophisticated expectations of their users and to continue to provide quality service and information to their users.

Gill Needham
Associate Director
The Open University Library

Index